MEM30007A

2015

Select common engineering materials

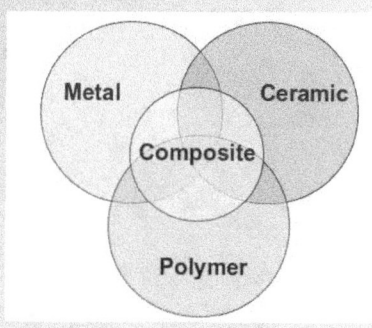

MEM30007A - Select common engineering materials

Feedback:

Your feedback is essential for improving the quality of these manuals.

This unit has not been technically edited. Please advise BlackLine Design of any changes, additions, deletions or anything else you believe would improve the quality of this Student Workbook. Don't assume that someone else will do it. Your comments can be made by photocopying the relevant pages and including your comments or suggestions.

Forward your comments to:

>BlackLine Design
>blakline@bigpond.net.au
>Sydney, NSW 2000

Corporate Licenses

State and National TAFE Colleges and Institutes, and Registered Training Organisations are eligible to purchase corporate licenses.

All licenses are perpetual and allow the licensee to upload the material onto a delivery system (Moodle etc), print the resource in book form and sell or distribute the material to enrolled students within their organisation. The license allows the holder to re-badge the material but must retain acknowledgment to BlackLine Design as the original developer and owner.

First Published April 2013

This work is copyright. Any inquiries about the use of this material should be directed to the publisher.

Edition 1 – April 2013
Edition 2 – October 2015

MEM30007A - Select common engineering materials

Conditions of Use:

Unit Resource Manual

Manufacturing Skills Australia Courses

This Student's Manual has been developed by BlackLine Design Manufacturing Skills Australia Courses.

Additional resource units can viewed and be ordered at *www.acru.com*

All rights reserved. No part of this publication may be printed or t form by any means without the explicit permission of the writer.

Statutory copyright restrictions apply to this material in digital

Copyright © BlackLine Design 2015

BlackLine Design
1st October 2015 – Edition 2

MEM30007A - Select common engineering materials

Aims of the Competency Unit:

This unit covers recognising common materials used in engineering, assisting in the selection of a material for a specific application, and using test results to evaluate the properties of materials.

Unit Hours:
36 Hours

Prerequisites:

MEM30007A - Select common engineering materials

Elements and Performance Criteria

1. Identify common engineering materials by their principal properties
   - 1.1 The principal properties of ferrous and non-ferrous metals are identified.
   - 1.2 The principal properties of thermosetting and thermoplastic polymers are identified
   - 1.3 The principal properties of ceramics and composite materials are identified.
   - 1.4 The effects of different types of bonding in materials are identified.
   - 1.5 The effects of mechanical and thermal processes on the principal properties of materials are identified.

2. Select materials for specific applications
   - 2.1 The engineering requirement for the specific application is determined in consultation with others.
   - 2.2 Material is selected based on the requirement and consideration of principal properties and further processing.
   - 2.3 Selection is confirmed according to standard operating procedures.

3. Verify selected material as fit for purpose
   - 3.1 Appropriate tests for the required properties are identified.
   - 3.2 Testing of materials is arranged with appropriate persons, if necessary.
   - 3.3 Test results are analysed and material choices are confirmed or modified as appropriate.

BlackLine Design
1st October 2015 – Edition 2

MEM30007A - Select common engineering materials

Required Skills and Knowledge

Required skills include:

- undertaking research
- selecting/carrying out tests appropriate to the material
- communicating
- documenting
- planning and sequencing operations
- reading, interpreting and following information on written job instructions, specifications, standard operating procedures, charts, lists, drawings and other applicable reference documents

Required knowledge includes:

- classification of materials:
    - metals and non-metals
    - ferrous and non-ferrous metals
    - polymers (thermoplastics, thermosetting and elastomers)
    - ceramics
    - composite materials
- structure of materials
- physical properties of materials:
    - electrical conductivity/resistivity
    - specific gravity/density
    - thermal conductivity/expansion
    - specific heat
    - melting/boiling points
- magnetic properties
- optical properties
- mechanical properties:
    - strength - yield, tensile, compressive
    - stress/strain data
    - hardness
    - toughness (impact and slow strain)
    - elasticity
    - plasticity
    - ductility
    - malleability
    - fatigue
    - creep
- chemical properties:
    - corrosion of metals, corrosion processes, mechanisms
    - degradation of polymers
- materials testing methods - destructive testing and applications:
    - tensile
    - compressive
    - shear
    - torsion
    - hardness
    - impact
    - fatigue
    - creep
    - visual
    - corrosion testing
- engineering materials
- engineering applications of ferrous metals:
    - cast irons
    - carbon and alloy steels

- stainless steels
- engineering applications of non-ferrous metals:
  - aluminium and its alloys
  - copper, brass and bronze
  - nickel alloys, zinc, titanium
  - magnesium
  - refractory metals
- engineering applications of polymers:
  - thermosetting polymers
  - thermoplastic polymers
  - ceramics and glasses
- effects of mechanical and thermal processes on the properties of materials:
  - casting
  - forging, rolling and extrusion
  - cold forming
  - powder processes
  - heat treatment
  - joining - fasteners
  - soldering
  - brazing
  - welding
  - adhesives
  - finishing - coatings, metallic and non-metallic
- hazards and control measure associated with selecting common engineering materials, including housekeeping
- safe work practices and procedures

MEM30007A - Select common engineering materials

Lesson Program:

Unit hour unit and is divided into the following program.

Topic	Review Questions
Topic 1 – Properties of Materials:	MEM30007-RQ-01
Topic 2 – Properties Data:	MEM30007-RQ-02
Topic 3 – Materials Testing:	MEM30007-RQ-03
Topic 4 – Structure and Properties:	MEM30007-RQ-04
Topic 5 – Processing of Materials:	MEM30007-RQ-05
Topic 6 – Selection of Materials:	MEM30007-RQ-06
Topic 7 – Safety Parameters:	MEM30007-RQ-07

MEM30007A - Select common engineering materials

Contents:

Conditions of Use: 3
Unit Resource Manual 3
Manufacturing Skills Australia Courses 3
Feedback: 4
Aims of the Competency Unit: 5
Unit Hours: 5
Prerequisites: 5
Elements and Performance Criteria 6
Required Skills and Knowledge 7
Lesson Program: 9
Contents: 10
Terminology: 13

Materials: 20
Engineering Metals: 20
Engineering Polymers : 22
Engineering Ceramics: 25

Topic 1 – Properties of Materials: 27
Required Skills: 27
Required Knowledge: 27
Introduction to Selection of Materials: 27
Properties of materials 28
Mechanical Properties: 29
   Strength: *30*
   Stiffness: *32*
   Ductility or Brittleness: *33*
   Toughness: *34*
   Hardness: *35*
Electrical Properties: 36
Thermal Properties: 36
Physical Properties: 37
Chemical Properties: 38
The Range of Materials: 38
   Metals: *38*
   Polymers and Elastomers: *39*
   Ceramics and Glasses: *39*
   Composites: *39*
Costs: 40
Review Problems: 42

Topic 2 – Properties Data: 44
Required Skills: 44
Required Knowledge: 44
Standards: 44
Data Sources: 44
Coding Systems: 45
   Steel: *45*
   Stainless Steel: *46*
   Aluminium: *46*
   Copper: *47*
   Plastics: *48*
   Timber: *48*
   Alloy Steel: *48*
Data Analysis: 53
Review Problems: 55

MEM30007A - Select common engineering materials

Topic 3 – Materials Testing: 56
    Required Skills: 56
    Required Knowledge: 56
    Standard Tests: 56
    The Tensile Test: 56
        The Test Piece: 56
        Tensile Test Results: 58
        Validity of Tensile Test Data: 59
        Interpreting Tensile Test Data: 59
        Tensile Tests for Plastics: 60
    Bend Tests: 61
    Impact Tests: 62
        Izod V-Notch Test Pieces: 62
        Charpy V-Notch Test Pieces: 63
    Impact Test Results: 64
        Interpreting Impact Test Results: 64
    Toughness Test: 65
    Hardness Tests: 65
        The Brinell Hardness Test: 65
        The Vickers Hardness Test: 66
        The Rockwell Hardness Test: 67
        Comparison of the Different Hardness Scales: 68
        The Moh Scale of Hardness: 68
        Hardness Values: 69
    Review Problems: 70

Topic 4 – Structure and Properties: 73
    Required Skills: 73
    Required Knowledge: 73
    Structure of Metals: 73
        Crystals: 74
        Crystalline Structure: 75
    Alloys: 76
        Ferrous Alloys: 77
        Plain Carbon Steel: 79
        Non-Ferrous Alloys: 80
    Stretching Metals: 82
    Cold Working: 85
        Heat Treating Cold-Worked Metals: 86
    Hot Working: 89
    The Structure of Polymers: 89
        Additives: 90
    Thermoplastics: 90
        Examples of Thermoplastics: 92
    Thermosets: 96
        Examples of Thermosets: 96
    Elastomers: 98
        Examples of Elastomers: 98
    The Structure of Composites: 99
        Fibres in a Matrix: 100
    Electrical Conductivity: 101
    Review Problems: 103

Topic 5 – Processing of Materials: 105
    Required Skills: 105
    Required Knowledge: 105
    Shaping Metals: 105
        Casting: 105
        Manipulative processes: 107
        Powder Techniques: 108

MEM30007A - Select common engineering materials

Shaping Polymers: 108
   Flowing Processes: *109*
   Manipulative Processes: *110*
   Drawing of Polymers: *110*
Heat Treatment of Metals: 111
   Surface Hardening: *113*
Integrated Circuit Fabrication: 114
Review Problems: 115

Topic 6 – Selection of Materials: 117
Required Skills: 117
Required Knowledge: 117
Requirements: 117
   Stages in the selection process *118*
Costs: 118
Failure in Service: 119
   The causes of Failure: *119*
   Examination of Failures: *120*
Selection of Materials: 122
   Car Bodywork: *122*
   Tennis Racket: *123*
   Small Components for Toys: *124*
Review Problems: 126

Topic 7 – Safety Parameters: 127
Required Skills: 127
Required Knowledge: 127
Health and safety at work 127
   Employer's Responsibilities *127*
   Employee's Responsibilities: *128*
   Government Responsibilities: *128*
   Trade Union Responsibilities: *129*
   Designers, Manufacturers, Importers and Suppliers Responsibilities: *129*
   Industry Association Responsibilities: *129*
Safe Work Systems: 129
   Protective Clothing and Equipment: *130*
   Accidents and Emergencies: *130*
A Safe and Healthy Environment: 131
Review Problems: 132

Answers: 133
   Topic 1: *133*
   Topic 2: *133*
   Topic 3: *133*
   Topic 4: *134*
   Topic 5: *134*
   Topic 6: *135*
   Topic 7: *135*

Terminology:

Additives	Plastics and rubbers almost invariably contain, in addition to the polymer or polymers, other materials, i.e. Additives. These are added to modify the properties and cost of the material
Ageing	This term is used to describe a change in properties that occurs with certain metals due to precipitation occurring, there being no change in chemical composition
Alloy	A metal which is a mixture of two or more elements.
Amorphous	An amorphous material is a non-crystalline material, i.e. It has a structure which is not orderly
Annealing	This involves heating to and holding at a temperature which is high enough for recrystallization to occur and which results in a softened state for a material after a suitable rate of cooling, generally slowly. The purpose of annealing can be to facilitate cold working, improve machinability and mechanical properties, etc.
Anodizing	Describes the process, generally with aluminium, whereby a protective coating is produced on the surface of the metal by converting it to an oxide.
Austenite	Describes the structure of a face-centred cubic iron crystalline structure which has carbon atoms in the gaps in the face-centred iron.
Bend, angle of	The results of a bend test on a material are specified in terms of the angle through which the material can be bent without breaking. The greater the angle, the more ductile the material.
Brinell number	The Brinell number is the number given to a material as a result of a Brinell test and is a measure of the hardness of a material. The larger the number, the harder the material.
Brittle failure	With brittle failure a crack is initiated and propagates prior to any significant plastic deformation. The fracture surface of a metal with a brittle fracture is bright and granular due to the reflection of light from individual crystal surfaces. With polymeric materials the fracture surface may be smooth and glassy or somewhat splintered and irregular.
Brittle material	A brittle material shows little plastic deformation before fracture. The material used for a china teacup is brittle. Thus because there is little plastic reformation before breaking, a broken teacup can be stuck back together again to give the cup the same size and shape as the original.
Carburizing	A treatment which results in a hard surface layer being produced with ferrous alloys. The treatment involves heating the alloy in a carbon-rich atmosphere so that carbon diffuses into the surface layers, then quenching to convert the surface layers to martensite.
Case hardening	The term is used to describe processes in which, by changing the composition of surface layers of ferrous alloys, a hardened surface layer can be produced.
Casting	A manufacturing process which involves pouring liquid metal into a mould or, in the case of plastics, the mixing of the constituents in a mould.

MEM30007A - Select common engineering materials
Terminology

Cementite	A compound formed of iron and carbon, often referred to as iron carbide. It is a hard and brittle material.
Charpy Test Value	The Charpy test is used to determine the response of a material to a high rate of loading and involves a test piece being struck a sudden blow. The results are expressed in terms of the amount of energy absorbed by the test piece when it breaks. The higher the test value, the more ductile the material.
Cold Working	This is when a metal is subject to working at a temperature below its recrystallization temperature.
Composite	A material composed of two different materials bonded together in such a way that one serves as the matrix surrounding fibres or particles of the other.
Compressive Strength	The compressive strength is the maximum compressive stress a material can withstand before fracture.
Copolymer	A polymeric material produced by combining two or more monomers into a single polymer chain.
Corrosion Resistance	The ability of a material to resist deterioration by reacting with its immediate environment. There are many forms of corrosion and so there is no unique way of specifying the corrosion resistance of a material.
Creep	Creep is the continuing deformation of a material with the passage of time when it is subject to a constant stress. For a particular material the creep behaviour depends on both the temperature and the initial stress, the behaviour also depending on the material concerned.
Crystalline	Describes a structure in which there is a regular and orderly arrangement of atoms or molecules.
Damping Capacity	The damping capacity is an indicator of the ability of a material to suppress vibrations.
Density	Density is mass per unit volume.
Dielectric Strength	The dielectric strength is a measure of the highest potential difference an insulating material can withstand without electric breakdown. It is the breakdown voltage divided by the thickness of the material.
Ductile Failure	With ductile failure there is a considerable amount of plastic deformation prior to failure. With metals the fracture shows a typical cone and cup formation and the fracture surfaces are rough and fibrous in appearance.
Ductile Materials	Ductile materials show a considerable amount of plastic deformation before breaking. Such materials have a large value of percentage elongation.
Elastic Limit	The elastic limit is the maximum force or stress at which on its removal the material returns to its original dimensions.
Electrical Conductance	The reciprocal of the electrical resistance and has the unit of the Siemen (S). It is thus the current through a material divided by the voltage across it.
Electrical Conductivity	The electrical conductivity is defined by: $$\text{Conductivity} = \frac{L}{}$$

	RA
	Where R is the resistance of a strip of the material of length L and cross-sectional area A. Conductivity has the unit of S/m. The IACS specification of conductivity is based on 100% corresponding to the conductivity of annealed copper at 20°C and all other materials are then expressed as a percentage of this value.
Electrical Resistance	The current through it, the unit being the ohm (**W**).
Electrical resistivity	The electrical resistivity is defined by: $$\text{Resistivity} = \frac{RA}{L}$$ Resistivity has the unit **W** m
Expansion, coefficient of linear	The coefficient of linear expansion is a measure of the amount by which a unit length of a material will expand when the temperature rises by one degree. It is defined by: $$\text{Coefficient} = \frac{\text{change in length}}{\text{length x temp, change}}$$ It has the unit °C^{-1} or K^{-1}.
Expansivity, linear	This is an alternative name for the coefficient of linear expansion.
Fatigue life	The fatigue life is the number of stress cycles to cause failure.
Ferrite	The term is usually used for a structure consisting of carbon atoms lodged in body-centred cubic iron. Ferrite is comparatively soft and ductile.
Fracture toughness	The plane strain fracture toughness is an indicator of whether a crack will grow or not and thus is a measure of the toughness of a material when there is a crack present
Full hard	The term is used to describe the temper of alloys. It corresponds to the cold-worked condition beyond which the material can no longer be worked.
Grain	The term is used for a crystalline region within a metal, i.e. a region of orderly packed atoms.
Half-hard	The term is used to describe the temper of alloys. It corresponds to the cold-worked condition half-way between soft and full hard.
Hardness	The hardness of a material may be specified in terms of some standard test involving indentation, e.g. the Brinell, Vickers and Rockwell tests, or scratching of the surface of the material, the Moh test.
Heat treatment	This term is used to describe the controlled heating and cooling of metals in the sold state for the purpose of altering their properties.
Hooke's law	When a material obeys Hooke's law its extension is directly proportional to the applied stretching forces.
Hot working	This is when a metal is subject to working at a temperature in excess of its recrystallization temperature.
Impact properties	See *Charpy test value* and *Izod test value.*
Izod test value	The Izod test is used to determine the response of a material to a high rate of loading and involves a test piece being struck a sudden

	blow. The results are expressed in terms of the amount of energy absorbed by the test piece when it breaks. The higher the test value, the more ductile the material.
Limit of proportionality	Up to the limit of proportionality the extension is directly proportional to the applied stretching forces, i.e. the strain is proportional to the applied stress.
Malleability	Describes the ability of metals to permit plastic deformation in compression without rupturing.
Martensite	A general term used to describe a form of structure. In the case of ferrous alloys it is a structure produced when the rate of cooling from the austenitic state is too rapid to allow carbon atoms to diffuse out of the face-centred cubic form of austenite and produce the body-centred form of ferrite. The result is a highly strained hard structure.
Melting point	The temperature at which a material changes from solid to liquid.
Moh scale	A scale of hardness arrived at when considering the ease of scratching a material. It is a scale of 10, with the higher the number, the harder the material.
Monomer	The unit consisting of a relatively few atoms which are joined together in large numbers to form a polymer.
Nitriding	A treatment in which nitrogen diffuses into surface layers of a ferrous alloy and hard nitrides are produced with a hard surface layer.
Orientation	A polymeric material is said to have an orientation, uniaxial or biaxial, if during the processing of the material the molecules become aligned in particular directions. The properties of the material in such directions are markedly different from those in other directions.
Pearlite	A lamellar structure of ferrite and cementite.
Percentage elongation	The percentage elongation is a measure of the ductility of a material, the higher the percentage, the greater the ductility. It is the change in length which has occurred during a tensile test to breaking expressed as a percentage of the original length: $$\% \text{ Elongation} = \frac{\text{final - initial length}}{\text{initial length}} \times 100$$
Percentage reduction in area	A measure of the ductility of a material and is the change in cross-sectional area which has occurred during a tensile test to breaking expressed as a percentage of the original cross-sectional area.
Precipitation hardening	A heat treatment process which results in a precipitate being produced in such a way that a harder material is produced.
Proof stress	The 0.2% proof stress is defined as that stress which results in a 0.2% offset, i.e. the stress given by a line drawn on the stress-strain graph parallel to the linear part of the graph and passing through the 0.2% strain value. The 0.1% proof stress is similarly defined. Proof stresses are quoted when a material has no well-defined yield point.
Quenching	The method used to produce rapid cooling. In the case of ferrous alloys it involves cooling from the austenitic state by immersion in cold water or an oil bath.

MEM30007A - Select common engineering materials
Terminology

Recovery	The term is used for the treatment involving the heating of a metal so as to reduce internal stresses.
Recrystallization	Generally used to describe the process whereby a new, strain-free grain structure is produced from that existing in a cold-worked metal by heating.
Resilience	The term is used with elastomers to give a measure of the 'elasticity' of a material. A high-resilience material will suffer elastic collisions when a high percentage of the kinetic energy before the collision is returned to the object after it. A less resilient material would lose more kinetic energy in the collision.
Rockwell Test Value	The Rockwell test is used to give a value for the hardness of a material. There are a number of Rockwell scales and thus the scale being used must be quoted with all test results.
Ruling Section	The limiting ruling section is the maximum diameter of round bar at the centre of which the specified properties may be obtained.
Secant Modulus	For many polymeric materials there is no linear part of the stress-strain graph and thus a tensile modulus cannot be quoted. In such cases the secant modulus is used. It is the stress at a value of 0.2% strain divided by that strain.
Shear	When a material is loaded in such a way that one layer of the material is made to slide over an adjacent layer then the material is said to be in shear.
Shear Strength	The shear strength is the shear stress required to produce fracture.
Sintering	The process by which powders are bonded by molecular or atomic attraction as a result of heating to a temperature below the melting points of the constituent powders.
Solution Treatment	This heat treatment involves heating an alloy to a suitable temperature, holding at that temperature long enough for one or more constituent elements to enter into the crystalline structure, and then cooling rapidly enough for these to remain in solid solution.
Specific Gravity	The specific gravity of a material is the ratio of its density compared with that of water.
Specific Heat Capacity	The amount by which the temperature rises for a material when there is a heat input depends on its specific heat capacity. The higher the specific heat capacity, the smaller the rise in temperature per unit mass for a given heat input: $$\text{Specific heat capacity} = \frac{\text{heat input}}{\text{mass x temp, change}}$$ Specific heat capacity has the unit $J\ kg^{-1}\ K^{-1}$.
Specific Stiffness	The modulus of elasticity divided by the density.
Specific Strength	The strength divided by the density.
Stiffness	The property is described by the modulus of elasticity.
Strain	The property is described by the modulus of elasticity. Strain The engineering strain is defined as the ratio (change in length)/(original length) when a material is subject to tensile or compressive forces. Shear strain is the ratio (amount by which one layer slides over

	another)/(separation of the layers). Because it is a ratio, strain has no units, though it is often expressed as a percentage. Shear strain is usually quoted as an angle in radians.
Strength	See *Compressive strength. Shear strength* and *Tensile strength.*
Stress	In engineering, tensile and compressive stress is usually defined as (force)/(initial cross-sectional area). The true stress is (force)/(cross-sectional area at that force). Shear stress is the (shear force)/(area resisting shear). Stress has the unit Pa (Pascal) with $1\ Pa = 1\ N\ m^{-2}$.
Stress Relieving	A treatment to reduce residual stresses by heating the material to a suitable temperature, followed by slow cooling.
Stress-Strain Graph	The stress-strain graph is usually drawn using the engineering stress (see *Stress)* and engineering strain (see *Strain).*
Surface Hardening	A general term used to describe a range of processes by which the surface of a ferrous alloy is made harder than its core.
Temper	Used with non-ferrous alloys as an indication of the degree of hardness/strength, with expressions such as hard, half-hard, three-quarters hard being used.
Tempering	The heating of a previously quenched material to produce an increase in ductility.
Tensile Modulus	The tensile modulus, or Young's modulus, is the slope of the stress-strain graph over its initial straight-line region.
Tensile Strength	defined as the maximum tensile stress a material can withstand before breaking.
	$$\text{Thermal conductivity} = \frac{\text{rate of transfer of heat}}{\text{area x temp, gradient}}$$ Thermal conductivity has the unit $W\ m^{-2}\ K^{-1}$.
Thermal Expansivity	See *Expansion, coefficient of linear*.
Toughness	This property describes the ability of a material to absorb energy and deform plastically without fracturing. It is usually measured with the Izod test or the Charpy test. Another form of measure is the fracture toughness. See *Fracture toughness*.
Transition temperature	The transition temperature is the temperature at which a material changes from giving a ductile failure to giving a brittle failure.
Vickers Test Results	The Vickers test is used to give measure of hardness. The higher the Vickers hardness number, the greater the hardness.
Water Absorption	The percentage gain in weight of a polymeric material after immersion in water for a specified amount of time under controlled conditions.
Wear Resistance	A subjective comparison of the wear resistance of materials. There is no standard test.
Work Hardening	The hardening of a material produced as a consequence of working subjecting it to plastic deformation at temperatures below those of recrystallization.

Yield Point	For many metals, when the stretching forces applied to a test piece are steadily increased a point is reached when the extension is no longer proportional to the applied forces and the extension increases more rapidly than the force until a maximum force is reached; this is called the upper yield point. The force then drops to a value called the lower yield point before increasing again as the extension is continued.
Young's Modulus	See *Tensile modulus*.

Materials:

Engineering Metals:
The following is an alphabetical listing of metals, each being listed according to the main alloying element, with their key characteristics. It is not a comprehensive list of all metallic elements, just those commonly encountered in engineering.

Aluminium
Used in commercially pure form and alloyed with copper, manganese, silicon, magnesium, tin and zinc. Alloys exist in two groups; casting alloys and wrought alloys. Some alloys can be heat treated. Aluminium and its alloys have a low density, high electrical and thermal conductivity and excellent corrosion resistance. Tensile strength tends to be of the order of 150 to 400 MPa with the tensile modulus about 70 GPa; there is a high strength-to-weight ratio.

Chromium
Chromium is mainly used as an alloying element in stainless steels, heat-resistant alloys and high-strength alloy steels. It is generally used in these for the corrosion and oxidation resistance it confers on the alloys.

Cobalt
Cobalt is widely used as an alloy for magnets, typically 5-35% cobalt with 14-30% nickel and 6-13% aluminium. Cobalt is also used for alloys which have high strength and hardness at room and high temperatures; these are often referred to as Stellates. Cobalt is also used as an alloying element in steels.

Copper
Copper is very widely used in the commercially pure form and alloyed in the form of brasses, bronzes, cupro-nickels and nickel silvers. Brasses are copper-zinc alloys containing up to 43% zinc. Bronzes are copper-tin alloys. Copper-aluminium alloys are referred to as aluminium bronzes, copper-silicon alloys as silicon bronzes. Copper-beryllium alloys as beryllium bronzes. Cupro-nickels are copper-nickel alloys. Copper and its alloys have good corrosion resistance, high electrical and thermal conductivity, good machinability, can be joined by soldering, brazing and welding, and generally have good properties at low temperatures. The alloys have tensile strengths ranging from about 180 to 300 MPa and a tensile modulus about 20 to 28 GPa.

Gold
Gold is very ductile and readily cold worked. It has good electrical and thermal conductivity.

Iron
The term ferrous alloy is used for the alloys of iron; these alloys include carbon steels, cast irons, alloy steels and stainless steels. Steels have 0.05-2% carbon, cast irons 2 - 4.3% carbon. The term carbon steel is used for those steels in which essentially just iron and carbon are present. Steels with between 0.10% and 0.25% are termed mild steels, between 0.20% and 0.50% medium-carbon steels and 0.50-2% carbon as high-carbon steels. With such steels in the annealed state the tensile strength increases from about 250 MPa at low carbon content to 900 MPa at high carbon content, the higher the carbon content, the more brittle the alloy. The term low-alloy steel is used for alloy steels when the alloying additions are less than 2%, medium-alloy between 2% and 10% and high-alloy when over 10%. In all cases the carbon content is less that 1%. Examples of low-alloy steels are manganese steels with strengths of the order of 500 MPa in the annealed state and 700 MPa when quenched and tempered. Stainless steels are high-alloy steels with more than 12% chromium. The modulus of elasticity of steels tends to be about 200 to 207 GPa.

Lead
Other than its use in lead storage batteries, it finds a use in lead-tin alloys as a metal solder and in steels to improve the machinability.

Magnesium
Magnesium is used in engineering alloyed mainly with aluminium, zinc and manganese. The alloys have a very low density and though tensile strengths are only of the order of 250 MPa there is a high strength-to-weight ratio. The alloys have a low tensile modulus, about 40 GPa. They have good machinability.

Molybdenum
Molybdenum has a high density, high electrical and thermal conductivity and low thermal expansivity. At high temperatures it oxidizes; it is used for electrodes and support members in electronic tubes and light bulbs, and heating elements for furnaces. Molybdenum is, however, more widely used as an alloying element in steels. In tool steels it improves hardness, in stainless steels it improves corrosion resistance, and in general in steels it improves strength, toughness and wear resistance.

Nickel
Nickel is used as the base metal for a number of alloys with excellent corrosion resistance and strength at high temperatures. The alloys are basically nickel-copper and nickel-chromium-iron. The alloys have tensile strengths between about 350 and 1400 MPa, the tensile modulus being about 220 GPa.

Niobium
Niobium has a high melting point, good oxidation resistance and low modulus of elasticity. Niobium alloys are used for high-temperature items in turbines and missiles. It is used as an alloying element in steels.

Palladium
Palladium is highly resistant to corrosion. It is alloyed with gold, silver or copper to give metals which are used mainly for electrical contacts.

Platinum
Platinum has a high resistance to corrosion, is very ductile and malleable, but expensive. It is widely used in jewellery. Alloyed with elements such as iridium and rhodium, the metal is used in instruments for items requiring a high resistance to corrosion.

Silver
Silver has a high thermal and electrical conductivity, and is very soft and ductile.

Tantalum
Tantalum is a high melting point, highly acid-resistant, very ductile metal. Tantalum-tungsten alloys have high melting points, high corrosion resistance and high tensile strength.

Tin
Tin has a low tensile strength, is fairly soft and can be very easily cut. Tin plate is steel plate coated with tin, the tin conferring good corrosion resistance. Children's toy soldiers were essentially manufactured from tin alloyed with lead and sometimes antimony. Tin alloyed with copper and antimony gives a material widely used for bearing surfaces.

Titanium
Titanium as a commercially pure or alloy has a high strength coupled with a relatively low density. It retains its properties over a wide temperature range and has excellent corrosion resistance. Tensile strengths are typically of the order of 1100 MPa and tensile modulus about 110 GPa.

Tungsten
Tungsten is a dense metal with the highest melting point of any metal (3410°C). It is used for light bulb and electronic tube filaments, electrical contacts, and as an alloying element in steels.

Zinc

Zinc has very good corrosion resistance and hence finds a use as a coating for steel, the product being called galvanized steel. It has a low melting point and hence zinc alloys are used for products such as small toys, cogs, shafts, door handles, etc. produced by die casting. Zinc alloys are generally about 96% zinc with 4% aluminium and small amounts of other elements or 95% zinc with 4% aluminium, 1% copper and small amounts of other elements. Such alloys have tensile strength of about 300 MPa, elongations of about 7-10% and hardness of about 90 BH.

Engineering Polymers :

The following is an alphabetical listing of the main polymers used in engineering, together with brief notes of their main characteristics.

Acrylonitrile-butadiene-styrene (ABS)

ABS is a thermoplastic polymer giving a range of opaque materials with good impact resistance, ductility and moderate tensile (17 to 58 MPa) and compressive strength. It has a reasonable tensile modulus (1.4 to 3.1 GPa) and hence stiffness, with good chemical resistance.

Acetal

Acetals (polyacetals), are thermoplastics with properties and applications similar to those of nylons. A high tensile strength (70 MPa) is retained in a wide range of environments; they have a high tensile modulus (3.6 GPa) and hence stiffness, high impact resistance and a low coefficient of friction. Ultraviolet radiation causes surface damage.

Acrylics

Acrylics are transparent thermoplastics, trade names for such materials including Perspex and Plexiglass; they have high tensile strength (50 to 70 MPa) and tensile modulus (2.7 to 3.5 GPa), hence stiffness, good impact resistance and chemical resistance, but a large thermal expansivity.

Butadiene-Acrylonitrile

This is an elastomer, generally referred to as nitrile or Buna-N rubber CNBR). It has excellent resistance to fuels and oils.

Butadiene-Styrene

Butadiene-styrene is an elastomer and is very widely used as a replacement for natural rubber because of its cheapness. It has good wear and weather resistance, good tensile strength, but poor resilience, poor fatigue strength and low resistance to fuels and oils.

Butyl

Buty (isobutene-isoprene copolymer) is an elastomer. It is extremely impermeable to gases.

Cellulosics

The term Cellulosics encompasses cellulose acetate, cellulose acetate butyrate, cellulose acetate propionate, cellulose nitrate and ethyl cellulose. All are thermoplastics. Cellulose acetate is a transparent material.; additives are required to improve toughness and heat resistance. Cellulose acetate butyrate is similar to cellulose acetate but less temperature sensitive and with a greater impact strength. It has a tensile strength of 18 to 48 MPa and a tensile modulus of 0.5 to 1.4 GPa. Cellulose nitrate colours and becomes brittle on exposure to sunlight. It also burns rapidly. Ethyl cellulose is tough and has low flammability.

Chlorosulphonated Polyethylene

Chlorosulphonated Polyethylene is an elastomer having excellent resistance to ozone with good chemical resistance, fatigue and impact properties.

Epoxies
Epoxy resins are, when cured, thermosets; they are frequently used with glass fibres to form composites. Such composites have high strength, of the order of 200 to 420 MPa, and stiffness, about 21 to 25 GPa.

Ethylene propylene
Ethylene propylene is an elastomer. The copolymer form, EPM , and the terpolymer form, EPDM , have very high resistance to oxygen, ozone and heat.

Ethylene vinyl acetate
Ethylene vinyl acetate is an elastomer which has good flexibility, impact strength and electrical insulation properties.

Fluorocarbons
These are polymers consisting of fluorine attached to carbon chains. See *Polytetrafluoroethylene*.

Fluorocarbons
Flurocarbons are polymers consisting of fluorine attached to carbon chains. See *Polytetrafluoroethylene*.

Fluorosilicones
Refer to *Silicone rubbers*.

Melamine Formaldehyde
The resin, a thermoset, is widely used for impregnating paper to form decorative panels, and as a laminate for table and kitchen unit surfaces. It is also used with fillers for moulding knobs, handles, etc. It has good chemical and water resistance, good colourability and good mechanical strength (55 to 85 MPa) and stiffness (7.0 to 10.5 GPa).

Natural Rubber
Natural rubber is an elastomer. It is inferior to synthetic rubbers in oil and solvent resistance and oxidation resistance and is subject to attack by ozone.

Nylons
The term nylon is used for a range of thermoplastic materials having the chemical name of polyamides. A numbering system is used to distinguish between the various forms, the most common engineering ones being nylon 6, nylon 6.6 and nylon 11. Nylons are translucent materials with high tensile strength and of medium stiffness. Tensile strengths are typically about 75 MPa and the tensile modulus about 1.1 to 3.3 GPa. Additives such as glass fibres are used to increase strength. Nylons have low coefficients of fiction, which can be further reduced by suitable additives; for this reason they are widely used for gears and rollers. All nylons absorb water.

Phenol formaldehyde
Phenol formaldehyde is a thermoset and is mainly used as a reinforced moulding powder. It is low cost, and has good heat resistance, dimensional stability, and water resistance. Unfilled it has a tensile strength of 35 to 55 MPa and a tensile modulus of 5.2 to 7.0 GPa.

Polyacetal
See *Acetals*.

Polyamides
See *Nylons*.

Polycarbonates
Polycarbonates are transparent thermoplastics with high impact strength, high tensile strength (55 to 65 MPa), high dimensional stability and good chemical resistance; they are moderately stiff (2.1 to 2.4 GPa) and have good heat resistance and can be used at temperatures up to 120°C.

Polychloroprene
Polychloroprene is usually called neoprene, is an elastomer. It has good resistance to oils and good weathering resistance.

Polyesters
Two forms of Polyesters are possible, thermoplastics and thermosets. Thermoplastic polyesters have good dimensional stability, excellent electrical resistivity and are tough; they discolour when subject to ultraviolet radiation. Thermoset polyesters are generally used with glass fibres to form composite materials.

Polyethylene
Polyethylene, or polythene, is a thermoplastic material. There are two main types: low density (LDPE) which has a branched polymer chain structure and high density (HDPE) with linear chains. Materials composed of blends of the two forms are available. LDPE has a fairly low tensile strength (8 to 16 MPa) and tensile modulus (0.1 to 0.3 GPa), with HDPE being stronger (22 to 38 MPa) and stiffer (0.4 to 1.3 GPa); both forms have good impermeability to gases and very low absorption rates for water.

Polyethylene Terephthalate (P E T)
PET is a thermoplastic polyester. It has good strength (50 to 70 MPa) and stiffness (2.1 to 4.4 GPa), is transparent and has good impermeability to gases. It is widely used as bottles for soft drinks and water.

Polypropylene
Polypropylene is a thermoplastic material with a low density, reasonable tensile strength (30 to 40 MPa) and stiffness (1.1 to 1.6 GPa) with properties similar to those of polyethylene. Additives are used to modify the properties.

Polypropylene Oxide
Polypropylene oxide is an elastomer with excellent impact and tear strengths, good resilience and good mechanical properties.

Polystyrene
Polystyrene is a transparent thermoplastic. It has moderate tensile strength (35 to 60 MPa), reasonable stiffness (2.5 to 4.1 GPa), but is fairly brittle and exposure to sunlight results in yellowing. It is attacked by many solvents. Toughened grades, when produced by blending with rubber, have better impact properties. Polystyrene has a strength of about 17 to 42 MPa and stiffness of 1.8 to 3.1 GPa.

Polysulphide
Polysulphide is an elastomer with excellent resistance to oils and solvents, and low permeability to gases. It can, however, be attacked by microorganisms.

Polysulphide
Polysulphide is a strong, comparatively stiff thermoplastic which can be used to a comparatively high temperature. I t has good dimensional stability and low creep with a strength of about 70 MPa and a stiffness of about 2.5 GPa.

Polytetrafluoroethylene (P T F E)
PTFE is a tough and flexible thermoplastic which can be used over a very wide temperature range. Because other materials will not bond with it, the material is used as a coating for items where non-stick facilities are required.

Polyvinyl
Polyvinyl is a group of thermoplastics and includes polyvinyl acetate, polyvinyl butyral, polyvinyl chloride (PVC), chlorinated polyvinyl chloride and vinyl copolymers. Polyvinyl acetate (PVA) is widely used in adhesives and paints. Polyvinyl butyral (PVB) is mainly used as a coating material or adhesive. PVC has high strength (52 to 58 MPa) and stiffness (2.4 to 3.1 GPa), being a rigid material. It is frequently combined with plasticizers to give a lower strength, less rigid, material. Chlorinated PVC is hard and rigid with excellent chemical and heat resistance. Vinyl copolymers can give a range of properties according to the constituents and their ratio. A common copolymer is vinyl

chloride with vinyl acetate in the ratio 85 to 15 and is a rigid material. A more flexible form has the ratio 95 to 5.

Silicone rubbers
Silicone rubbers or, as they are frequently called, fluoro-silicone rubbers have good resistance to oils, fuels and solvents at high and low temperatures; they do, however, have poor abrasion resistance.

Styrene-Butadiene Styrene
Styrene-butadiene-styrene is called a thermoplastic rubber. Its properties are controlled by the ratio of styrene to butadiene. The properties are comparable to those of natural rubber.

Urea Formaldehyde
Urea formaldehyde is a thermosetting material and has similar applications to melamine formaldehyde. Surface hardness is very good. The resin is also used as an adhesive.

Engineering Ceramics:
The term ceramics covers a wide range of materials and here only a few of the more commonly used engineering ceramics are considered.

Alumina
Alumina, i.e. aluminium oxide, is a ceramic which finds a wide variety of uses. It has excellent electrical insulation properties and resistance to hostile environments. Combined with silica it is used as refractory bricks.

Boron
Boron fibres are used as reinforcement in composites with materials such as nickel.

Boron Nitride
Boron nitride is a ceramic and is used as an electric insulator.

Carbides
A major use of ceramics is, when bonded with a metal binder to form a composite material, as cemented tips for tools. These are generally referred to as bonded carbides, the ceramics used being generally carbides of chromium, tantalum, titanium and tungsten.

Chromium Carbide
See *Carbides*.

Chromium Oxide
Chromium oxide is a ceramic and is used as a wear-resistant coating.

Glass
The basic ingredient of most glasses is silica, a ceramic. Glasses tend to have low ductility, a tensile strength which is markedly affected by microscopic defects and surface scratches, low thermal expansivity and conductivity (and hence poor resistance to thermal shock), good resistance to chemicals and good electrical insulation properties. Glass fibres are frequently used in composites with polymeric materials.

Kaolinite
Kaolinite is ceramic is a mixture of aluminium and silicon oxides, being a clay.

Magnesia
Magnesia, i.e. magnesium oxide, is a ceramic and is used to produce a brick called a dolomite refractory.

Pyrex
Pyrex is is a heat-resistant glass, being made with silica, limestone and boric oxide. See *Glasses*.

Silica
Silica forms the basis of a large variety of ceramics. It is, for example, combined with alumina to form refractory bricks and with magnesium ions to form asbestos. It is the basis of most glasses.

Silicon nitride
Silicon nitride is a ceramic is used as the fibre in reinforced materials such as epoxies.

Soda glass
Soda glass is the common window glass, being made from a mixture of silica, limestone and soda ash. See *Glasses*.

Tantalum Carbide
See Carbides.

Titanium Carbide
See Carbides.

Tungsten carbide
See Carbides

Topic 1 – Properties of Materials:

Required Skills:
On completion of the session, the participants will be able to:
- Recognise the link between the selection of materials for a product and the properties required of them by the product.
- Communicate in appropriate technical terms about the properties of materials.
- Recognise the properties characteristic of different groups of materials.

Required Knowledge:
- Processing formulae.
- Reading tables and charts.

Introduction to Selection of Materials:
The selection of a material for a component to be manufactured is as important as the design. The selection of a suitable material must be carefully thought out; information on the use of the component, working environment, wear, force and stress loading, aesthetics, compatibility of adjoining materials (galvanic corrosion) and financial constraints.

Material selection is a step in the process of designing any physical object. In the context of product design, the main goal of material selection is to minimize cost while meeting product performance goals. Systematic selection of the best material for a given application begins with properties and costs of candidate materials. For example, a thermal blanket must have poor thermal conductivity in order to minimize heat transfer for a given temperature difference.

Systematic selection for applications requiring multiple criteria is more complex. For example, a rod which should be stiff and light requires a material with high Young's modulus and low density. If the rod will be pulled in tension, the specific modulus, or modulus divided by density will determine the best material; because a plate's bending stiffness scales as its thickness cubed, the best material for a stiff and light plate is determined by the *cube root* of stiffness divided by density.

Question – What materials are used for a container of soft drink?
Answer – Soft drinks are manufactured in aluminium cans, glass or plastic bottles.

Question – What makes these materials suitable and others not?
Answer – In order to answer this question we need to investigate the properties of other materials.

Points to be considered in the selection of a material could be:
- Rigidity - the container does not stretch or become floppy under the content's weight.
- Strength – the container can stand the weight of the contents.
- Resistance to chemical attack from the contents.
- Retention of the gas component - prevent the gas from escaping through the container walls.
- Low density – so the container is not too heavy.
- Cost effectiveness – profit.
- Ease of manufacture – increases the profit.

The selection of a suitable material involves balancing a number of different specifications and making a choice of the material.

One of the best ways to collect research on materials is to carry out tests/experiments that will reflect their use in projects. For example, if the materials require for a project will need to be hardwearing, a test could be devised and tried out using several possible materials. The findings would be filed using a range of materials and mark/grade for each according to their resistance.

A range of tests could be devised to assess waterproof properties, impact resistance, flexibility rigidity, and many more.

The Rich Picture in Figure 1.1 shows the wide range of facts and issues relating to materials research that need to be considered when designing a product; this type of presentation could be the first page of materials research for a design project.

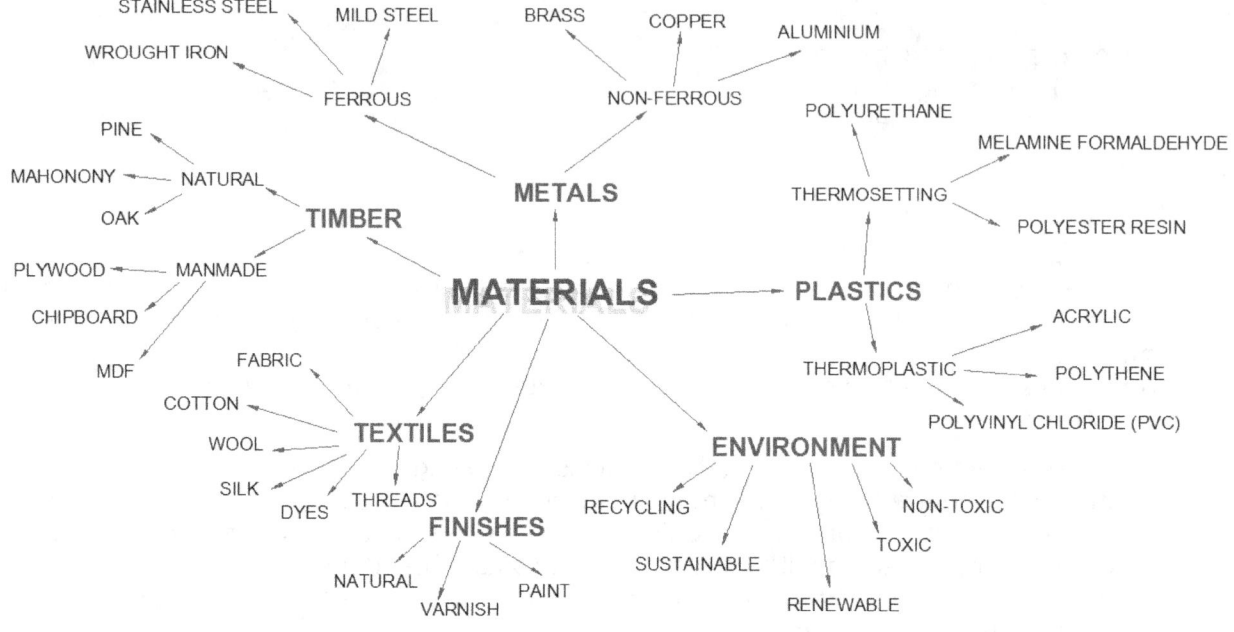

Figure 1.1

The selection of a material from which a product can be manufactured depends on a number of factors; these are often grouped under three main headings, namely:

1. The requirements imposed by the conditions under which the product is used, i.e. the service requirements; therefore, if a product is to be subject to forces then it might need strength, if subject to a corrosive environment then it might require corrosion resistance.

2. The requirements imposed by the methods proposed for the manufacture of the product. For example, if a material has to be bent as part of its processing, it must be ductile enough to be bent without breaking. A brittle material such as cast iron could not be used.

3. Cost.

Properties of materials

Materials selection for a product is based upon a consideration of the properties required including:

1. Mechanical properties –displayed when a force is applied to a material and include strength, stiffness, hardness, toughness and ductility.

2. Electrical properties –seen when the material is used in electrical circuits or components and include resistivity, conductivity and resistance to electrical breakdown.

3. Magnetic properties – relevant when the material is used as a magnet or part of an electrical component such as an inductor which relies on such properties.
4. Thermal properties – displayed when there is a heat input to a material and include expansivity and heat capacity.
5. Physical properties – the properties which are characteristic of a material and determined by its nature, including density, colour, surface texture.
6. Chemical properties - relevant in considerations of corrosion and solvent resistance.

The properties of materials are often changed markedly by the treatments they undergo; for example, steels can have their properties changed by heat treatment, such as annealing, which involves heating to some temperature and slowly cooling or quenching, i.e. heating and then immersing the material in cold water. Steel can also have its properties changed by working, for example, if a piece of carbon steel is permanently deform, it will have different mechanical properties from those existing before that deformation; refer to Topic 4 – Structure and Properties: for more information.

Mechanical Properties:

The mechanical properties are about the behaviour of materials when subject to forces. When a material is subject to external forces, then internal forces are set up in the materials which oppose the external forces. The material can be considered to be similar to a spring. A spring, when stretched by external forces, sets up internal opposing forces which are readily apparent when the spring is released and they force it to contract. When a material is subject to external forces which stretch it then it is in *tension* (Figure 1.2); when a material is subject to forces which squeeze it then it is in *compression* (Figure 1.3). If a material is subject to forces which cause it to twist or one face to slide relative to an opposite face then it is said to be in *shear* (Figure 1.4).

In discussing the application of forces to materials an important aspect is often not so much the size of the force as the force applied per unit area; thus, for example, if a strip of material is stretched by a force F applied over its cross-sectional area A, then the force applied per unit area is F/A. The term *stress* is used for the force per unit area.

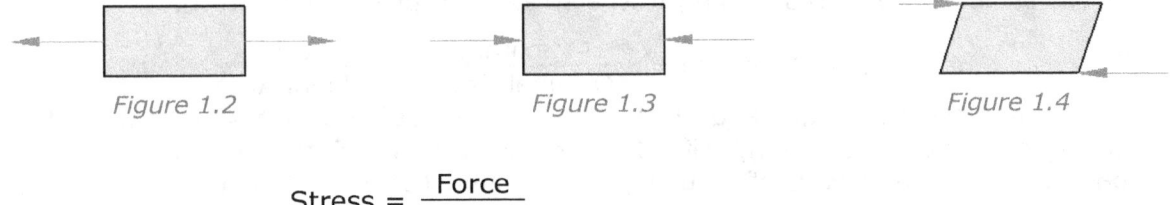

Figure 1.2   *Figure 1.3*   *Figure 1.4*

$$\text{Stress} = \frac{\text{Force}}{\text{Area}}$$

Stress has the units of pascal (Pa), with 1 Pa being a force of 1 newton per square metre, i.e. 1 Pa = 1 N/m². The stress is said to be direct stress when the area being stressed is at right angles to the line of action of the external forces, as when the material is in tension or compression. Shear stresses are not direct stresses since the forces being applied are in the same plane as the area being stressed. The area used in calculations of the stress is generally the original area that existed before the application of the forces. The stress is thus sometimes referred to as the engineering stress, the term true stress being used for the force divided by the actual area existing in the stressed state.

When a material is subject to tensile or compressive forces it changes in length. The formula to determine strain is:

$$\text{Strain} = \frac{\text{Change in Length}}{\text{Original Length}}$$

Since strain is a ratio of two lengths it has no units. Thus we might, for example, have a strain of 0.01. This would indicate that the change in length is 0.01 x the original length. However, strain is frequently expressed as a percentage.

$$\text{Strain as a \%} = \frac{\text{Change in Length}}{\text{Original Length}} \times 100$$

Therefore, the strain of 0.01 as a percentage is 1%, i.e. this is when the change in length is 1% of the original length.

Example:

A strip of material has a length of 50 mm. When it is subject to tensile forces it increases in length by 0.020 mm. What is the strain? The strain is the change in length divided by the original length:

$$\text{Strain} = \frac{0.020}{50} = 0.0004$$

Expressed as a percentage, the strain is:

$$\text{Strain} = \frac{0.020}{50} \times 100 = 0.04\%$$

Strength:

In materials science, the strength of a material is its ability to withstand an applied stress without failure. The field of strength of materials deals with loads, deformations and the forces acting on a material. A load applied to a mechanical member will induce internal forces within the member called stresses. The stresses acting on the material cause deformation of the material. Deformation of the material is called strain, while the intensity of the internal forces is called stress. The applied stress may be tensile, compressive, or shear. The strength of any material relies on three different types of analytical method: strength, stiffness and stability, where strength refers to the load carrying capacity, stiffness refers to the deformation or elongation, and stability refers to the ability to maintain its initial configuration. Material yield strength refers to the point on the engineering stress-strain curve (as opposed to true stress-strain curve) beyond which the material experiences deformations that will not be completely reversed upon removal of the loading. The ultimate strength refers to the point on the engineering stress-strain curve corresponding to the stress that produces fracture.

$$\text{Tensile Strength} = \frac{\text{Maximum Tensile Force}}{\text{Original Cross-Sectional Area}}$$

The compressive strength and shear strength are defined in a similar way. The unit of strength is the pascal (Pa), with 1 Pa being 1 N/m². Strengths are often millions of pascals and so the MPa is often used, 1 MPa being 106 Pa or 1 million Pa.

Often it is not the strength of a material that is important in determining the situations in which a material can be used but the value of the stress at which the material begins to yield. If gradually increasing tensile forces are applied to, say, a strip of mild steel then initially when the forces are released the material springs back to its original shape. The material is said to be *elastic*. If measurements are made of the extension at different forces and a graph plotted, then the extension is found to be proportional to the force and the material is said to obey *Hooke's law*. However, when a particular level of force is reached the material stops springing back completely to its original shape and is then said to show some plastic behaviour. This point coincides with the point on a force-extension graph at which the graph stops being a straight line graph, the so-called limit of proportionality.

Figure 1.5 – Force Extension Graph shows the type of force-extension graph which would be given by a sample of mild steel. The limit of proportionality is point A. Up to this point Hooke's law is obeyed and the material shows elastic behaviour, beyond it shows a mixture of elastic and plastic behaviour. Dividing the forces by the initial cross-sectional area of the sample and the extensions by the original length converts the force-extension data into a stress-strain graph, as in Figure 1.6 – Stress Strain Graph. The stress at which the material starts to behave in a non-elastic manner is called the elastic limit. Generally at almost the same stress the material begins to stretch without any further increase in force and is said to have yielded. The term yield stress is used for the stress at which this occurs; for some materials, such as mild steel, there are two yield points, termed the upper and the lower yield points. A carbon steel typically might have a tensile strength of 600 MPa and a yield stress of 300 MPa.

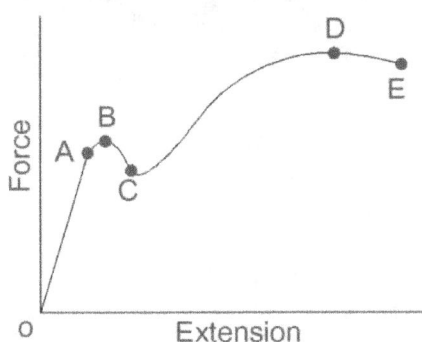

Figure 1.5 – Force Extension Graph

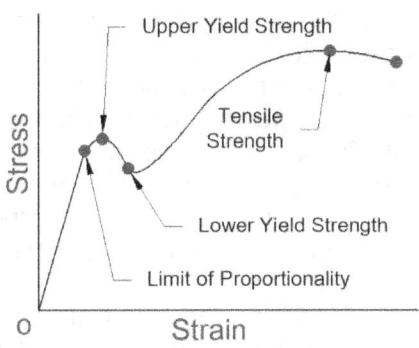

Figure 1.6 – Stress Strain Graph

Where:
- A = Limit of Proportionality
- B = Upper Yield Point
- C = Lower Yield Point
- D = Maximum Force
- E = Breaking Point

In some materials, such as aluminium alloys, the yield stress is not so easily identified as with mild steel and the term proof stress is used as a measure of when yielding begins; this is the stress at which the material has departed from the straight-line force-extension relationship by some specified amount. The 0.1% proof stress is defined as that stress which results in a 0.1% offset, i.e. the stress given by a line drawn on the stress-strain graph parallel to the linear part of the graph and passing through the 0.1% strain value, as in Figure 1.7 – Determination of Proof Stress. A 0.2% proof stress is likewise defined as that stress which results in a 0.2% offset.

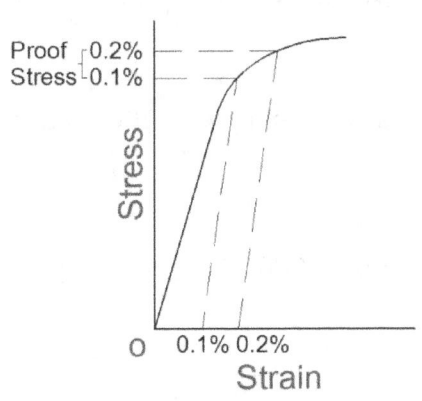

Figure 1.7 – Determination of Proof Stress

Since Stress is Force/Area then:

Yield Force = Yield Stress x Area
= $200 \times 10^6 \times 100 \times 10^{-6}$
= 20000N
or 20kN

Example:
Samples are taken of cast aluminium alloys and give the following data. Which is the strongest in tension?
- LM4 tensile strength 140 MPa
- LM6 tensile strength 160 MPa
- LM9 tensile strength 170 MPa

The strongest in tension is the one with the highest tensile strength and is LM9.

Stiffness:

The stiffness of a material is the ability of a material to resist bending; when a strip of material is bent, one surface is stretched and the opposite face is compressed, as illustrated in Figure 1.8.

The more a material bends, the greater is the amount by which the stretched surface extends and the compressed surface contracts. Therefore, a stiff material would be one that undergoes a small change in length when subject to such forces; this means a small strain when subject to such stress and so a small value of strain/stress, or conversely a large value of stress/strain. For most materials a graph of stress against strain gives initially a straight-line relationship, as illustrated in Figure 1.9. Thus a large value of stress/strain means a steep slope of the stress-strain graph. The quantity stress/strain when we are concerned with the straight-line part of the stress-strain graph is called the modulus of elasticity (or sometimes Young's modulus).

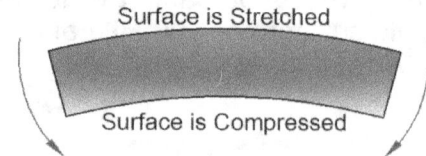

Figure 1.8 - Bending

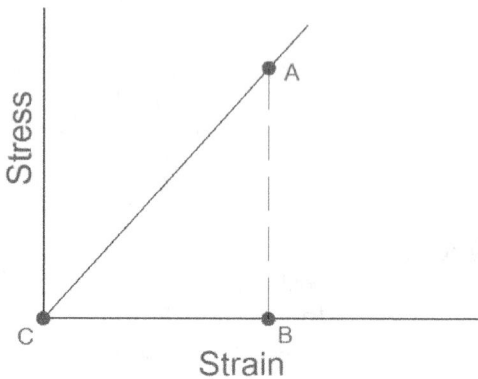

Figure 1.9 – Modulus of Elasticity = AB/BC

$$\text{Modulus of Elasticity} = \frac{\text{Stress}}{\text{Strain}}$$

The units of the modulus are the same as those of stress, since strain has no units. Engineering materials frequently have a modulus of the order of 1000 million Pa, i.e. 10^9 Pa; this is generally expressed as GPa, with 1 GPa = 10^9 Pa. Typical values are about 200 GPa for steels and about 70 GPa for aluminium alloys. A stiff material therefore has a high modulus of elasticity. Consequently steels are stiffer than aluminium alloys; for most engineering materials the modulus of elasticity is the same in tension as in compression.

Example:

If a material of a component has a tensile modulus of elasticity of 200 GPa, what strain will be produced by a stress of 4 MPa?

Since the modulus of elasticity is stress/strain then:

$$\text{Strain} = \frac{\text{Stress}}{\text{Modulus}} = \frac{4 \times 10^6}{200 \times 10^9} = 0.00002$$

Example:

Which of the following plastics is the stiffest?
- ABS   Tensile Modulus 2.5 GPa
- Polycarbonate   Tensile Modulus 2.8 GPa
- Polypropylene   Tensile Modulus 1.3 GPa
- PVC   Tensile Modulus 3.1 GPa

The stiffest plastic is the one with the highest tensile modulus and therefore is the PVC.

Ductility or Brittleness:

When a glass is dropped and breaks then it is possible (but not probable) to stick all the pieces together again and restore the glass to its original shape. The glass is said to be a **brittle** material.

Using the previous example of a soft drink container, if a steel or aluminium can is dropped, the can is less likely to shatter like a glass bottle but more likely to show permanent deformation in the form of dents. The material has shown plastic deformation which the term **permanent deformation** is used to changes in dimensions which are not removed when the forces applied to the material are taken away. Materials which develop significant permanent deformation before they break are called **ductile.** Figure 1.10 and Figure 1.11 show the types of stress-strain graphs given by brittle and ductile materials, the ductile one indicating a considerable extent of plastic behaviour.

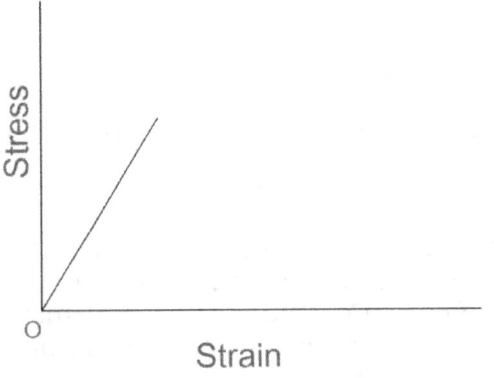

Figure 1.10 – Brittle Materials

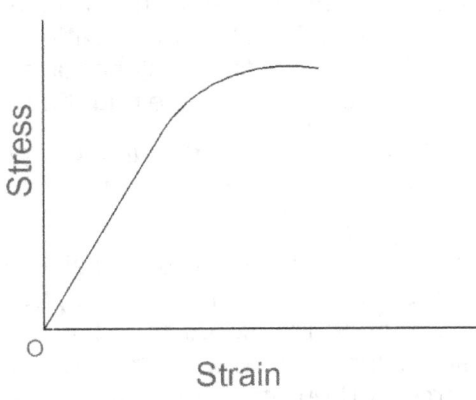

Figure 1.11 – Ductile Materials

A measure of the ductility of a material is obtained by determining the length of a test piece of the material, then stretching it until it breaks and then, by putting the pieces together, measuring the final length of the test piece, as illustrated in Figure 1.8. A brittle material will show little change in length from that of the original test piece, but a ductile material will indicate a significant increase in length. The measure of the ductility is then the **percentage elongation**.

$$\% \text{ elongation} = \frac{\text{Final} - \text{Initial Lengths}}{\text{Initial Length}} \times 100$$

A reasonably ductile material, such as mild steel, will have an elongation of about 20%, or more. A brittle material, such as a cast iron, will have an elongation of less than 1%

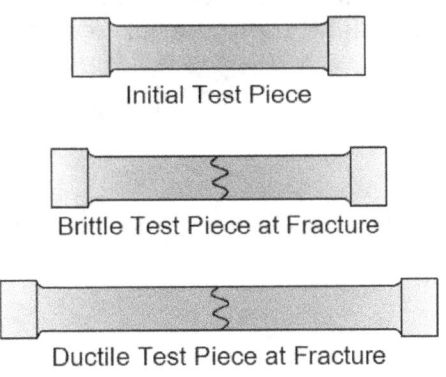

Figure 1.12 – Test Pieces after Fracture

Example

A material has an elongation of 10%. By how much longer will be a strip of the material of initial length 200 mm when it breaks? The percentage elongation can be expressed as

$$\% \text{ elongation} = \frac{\text{Change in Length}}{\text{Original Length}} \times 100$$

Therfore

$$\text{Changes in Length} = \frac{10 \times 200}{100} = 20 \text{ mm}$$

Example

Which of the following materials is the most ductile?
- 80-20 brass % elongation 50%
- 70-30 brass % elongation 70%
- 60-40 brass % elongation 40%

The most ductile material is the one with the largest percentage elongation, therefore the 70-30 brass.

Toughness:

A tough material can be considered to be one that resists breaking meaning that a tough material requires more energy to break it than a less tough one. There are, however, a number of measures that are used for toughness. Consider a length of material being stretched by tensile forces; when it is stretched by an amount as a result of a constant force F_1 then the work done is:

$$\text{Work} = \text{Force (F)} \times \text{extension (y)}$$

$$\text{Work} = F_1 y_1$$

Therefore, if a force-extension graph is considered as shown in Figure 1.13, the work done when a very small extension is considered, is the area of that strip under the graph. The total work done in stretching a material to an extension y_1, i.e. through an extension which we can consider to be made up of a number of small extensions, is thus:

$$\text{Work} = F_1 y_1 + F_2 y_2 + F_3 y_3 + \ldots$$

and so is the area under the graph up to x.

Since stress = force/area and strain - extension/length then:

$$\text{Work} = (\text{stress} \times \text{area}) \times (\text{strain} \times \text{length})$$

Since the product of the area and length is the volume of the material, then:

$$\text{Work/volume} = \text{stress} \times \text{strain}$$

Thus the work done in stretching a material unit volume to a particular strain is the sum of the work involved in stretching the material to each of the strains up to this strain.

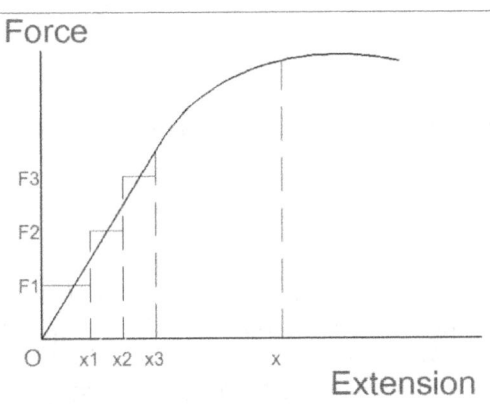

Figure 1.13

The area under a force-extension graph up to the breaking point is thus a measure of the energy required to break the material. The area under the stress-strain graph up to the breaking point is a measure of the energy required to break a unit volume of the material. A large area is given by a material with a large yield stress and high ductility (see Figure 1.10 and Figure 1.11). Such materials can thus be considered to be tough.

An alternative way of considering toughness is the ability of a material to withstand shock loads. A measure of this ability to withstand suddenly applied forces is obtained by impact tests, such as the Charpy and Izod tests (refer to Topic 3 – Materials Testing:). In these tests, a test piece is struck a sudden blow and the energy needed to break it is measured. The results are thus expressed in units of energy, i.e. joules (J). A brittle material will require less energy to break it than a ductile one. The results of such tests are often used as a measure of the brittleness of materials.

Another measure of toughness that can be used is fracture toughness. Fracture toughness can be defined as a measure of the ability of a material to resist the propagation of a crack. The toughness is determined by loading a sample of the material which contains a deliberately introduced crack of length 2c and recording the tensile stress a at which the crack propagates. The fracture toughness, symbol K_c and usual units MPa m^{½}, is given by:

$$K_c = sS\pi c$$

The smaller the value of the toughness, the more readily a crack propagates. The value of the toughness depends on the thickness of the material, high values occurring for thin sheets and decreasing with increasing thickness to become almost constant in thick sheets. For this reason, a value called the plane strain fracture toughness K_{Ic} is often quoted; this is the value of the toughness that would be obtained with thick sheets. Typical values are of the order of 1 MPa m^{½} for glass, which readily fractures when there is a crack present, to values of the order of 50 to 150 MPa m^{½} for some steels and copper alloys. In such materials cracks do not readily propagate.

Hardness:

The hardness of a material is a measure of its resistance to abrasion or indentation. A number of scales are used for hardness, depending on the method that has been used for measuring. The hardness of a material is a measure of how resistant solid matter is to various kinds of permanent shape change when a force is applied. Macroscopic hardness is generally characterized by strong intermolecular bonds, but the behavior of solid materials under force is complex; therefore, there are different measurements of hardness: scratch hardness, indentation hardness, and rebound hardness.

The hardness is roughly related to the tensile strength of a material, the tensile strength being roughly proportional to the hardness (refer to Topic 3 – Materials Testing:); therefore, the higher the hardness of a material, the higher is likely to be the tensile strength.

Hardness is dependent on ductility, elastic stiffness, plasticity, strain, strength, toughness, viscoelasticity, and viscosity.

Common examples of hard matter are ceramics, concrete, certain metals, and superhard materials, which can be contrasted with soft matter.

Electrical Properties:

The electrical **resistivity** (p) is a measure of the electrical resistance of a material, being defined by the equation:

$$P = \frac{RA}{L}$$

where R is the resistance of a length L of the material of cross-sectional area A. The unit of resistivity is the ohm metre. An electrical insulator such as a ceramic will have a very high resistivity, typically of the order of 10^{10} W m or higher. An electrical conductor such as copper will have a very low resistivity, typically of the order of 10^{10} W m.

The electrical conductance of a length of material is the reciprocal of its resistance and has the unit of Q^{-1}; this unit is given a special name of siemens (S). The electrical conductivity (a) is the reciprocal of the resistivity:

$$W = \frac{1}{W} = \frac{L}{RA}$$

The unit of conductivity is thus W^{-1} m or S m^{-1}. Since conductivity is the reciprocal of the resistivity, an electrical insulator will have a very low conductivity, of the order of 10^{-10} S/m, while an electrical conductor will have a very high one, of the order of 10^8 S/m.

The dielectric strength is a measure of the highest voltage that an insulating material can withstand without electrical breakdown. It is defined as:

$$\text{Dielectric Strength} \; \frac{\text{Breakdown Voltage}}{\text{Insulator Thickness}}$$

The units of dielectric strength are volts per metre. Polythene has a dielectric strength of about 4×10^7 V/m; this means that a 1 mm thickness of polythene will require a voltage of about 40 000 V across it before it will break down.

Example

An electrical capacitor is to be made with a sheet of polythene of thickness 0.1 mm between the capacitor plates. Determine the greatest voltage that can be connected between the capacitor plates if there is not to be electrical breakdown? Take the dielectric strength to be 4×10^7 V/m. The dielectric strength is defined as the breakdown voltage divided by the insulator thickness, hence:

$$\begin{aligned}
\text{Breakdown Voltage} &= \text{Dielectric Strength} \times \text{Thickness} \\
&= 4 \times 10^7 \times 0.1 \times 10^{-3} \\
&= 4000 \; V
\end{aligned}$$

Thermal Properties:

The SI unit of temperature is the kelvin with a temperature change of 1 K being the same as a change of 1°C. The kelvin is a unit of measurement for temperature. It is one of the seven base units in the International System of Units (SI) and is assigned the unit symbol K. The Kelvin scale is an absolute, thermodynamic temperature scale using as its null point absolute zero, the temperature at which all thermal motion ceases in the classical description of thermodynamics. The kelvin is defined as the fraction $1/273.16$ of the thermodynamic temperature of the triple point of water (exactly 0.01 °C or 32.018 °F).

The linear expansivity (a) or coefficient of linear expansion is a measure of the amount by which a length of material will expand when the temperature increases and is defined as:

$$a = \frac{\text{Change in Length}}{\text{Original Length} \times \text{Change in Temperature}}$$

It has the unit of K^{-1}.

The specific heat capacity (c) is a measure of the amount of heat needed to raise the temperature of the material. It is defined as:

$$c = \frac{\text{Amount of Heat}}{\text{Mass} \times \text{Change in Temperature}}$$

C has the unit of $J\ kg^{-1}\ K^{-1}$. Weight-for-weight metals require less heat to reach a particular temperature than plastics. This is because metals have smaller specific heat capacities. For example, copper has a specific heat capacity of about 340 $J\ kg^{-1}\ K^{-1}$ while polythene is about 1800 $J\ kg^{-1}\ K^{-1}$.

The thermal conductivity of a material is a measure of its ability to conduct heat. There will only be a net flow of heat energy through a length of material when there is a difference in temperature between the ends of the material. Thus the thermal conductivity is defined in terms of the quantity of heat that will flow per second through a temperature gradient.

$$\lambda = \frac{\text{Quantity of Heat/Second}}{\text{Temperature Gradient}}$$

λ has the unit of $W\ m^{-1}\ K^{-1}$. A high thermal conductivity means a good conductor of heat. Metals tend to be good conductors. For example, copper has a thermal conductivity of about 400 $W\ m^{-1}\ K^{-1}$. Materials which are poor conductors of heat have low thermal conductivities. For example, plastics have thermal conductivities of the order of 0.03 $W\ m^{-1}\ K^{-1}$.

Example
A designer of domestic pans requires a material for a handle which would enable a hot pan to be picked up with comfort, the handle not getting hot. What quantity should he or she look for in tables in order to find a suitable material?

What is required is a material with a low thermal conductivity, probably a small fraction of a $W\ m^{-1}\ K^{-1}$.

Physical Properties:
Physical properties are those that can be observed without changing the identity of the substance. The general properties of matter such as colour, density, hardness, are examples of physical properties. Properties that describe how a substance changes into a completely different substance are called chemical properties. Flammability and corrosion/oxidation resistance are examples of chemical properties.

The density (p) of a material is the mass per unit volume.

$$p = \frac{\text{Mass}}{\text{Volume}}$$

It has the unit of kg/m^3. It is often an important property that is required in addition to a mechanical property; thus, for example, an aircraft undercarriage is required to be not only strong but also of low mass. Therefore, what is required is as high a strength as possible with as low a density as possible, i.e. a high value of strength/density; this quantity is often referred to as the specific strength. Steels tend to have specific strengths of the order of 50 to 100 $MPa/Mg\ m^{-3}$ (note: 1 Mg is 10^6 g or 1000 kg), magnesium alloys about 140 $MPa/Mkg\ m^{-3}$ and titanium alloys about 250 $MPa/Mkg\ m^{-3}$. For example, a lower-strength magnesium alloy would be preferred to a higher-strength, but higher-density, steel.

Chemical Properties:

Chemical and physical properties may often be tabulated together in most handbooks. In general, the data associated with a compound contain name, empirical and structural formula, molecular weight, Chemical Abstract (CA) registry number, melting point, boiling point, density, color, solubility, oxidation or reduction potential, and various spectroscopic peaks. However, other literature must be consulted on chemical reactivity.

Chemical reactions usually involve the breakage and formation of some chemical bonds. All chemical reactions involve the redistribution of electrons among species involved. Chemical properties show the nature of its reactivity, the type of compounds and the category of reactions. Think of a compound, and classify it according to the following criteria.

Attack on materials by the environment in which they are situated is a major problem. The rusting of iron is an obvious example. Tables are often used giving the comparative resistance to attack of materials in various environments, e.g. in aerated water, in salt water, to strong acids, to strong alkalis, to organic solvents, to ultraviolet radiation. Thus, for example, in a salt water environment carbon steels are rated at having very poor resistance to attack, aluminium alloys good resistance and stainless steels excellent resistance.

The Range of Materials:

Materials are usually classified into four main groups, these being metals, polymers and elastomers, ceramics and glasses, and composites. The following is a brief comparison, in general, of the properties of these main groups. Differences in the internal structure of the groups are discussed in Topic 4 – Structure and Properties:

Property	Metals	Polymers	Ceramics
Density (Mg m^{-3})	2 – 16	1 – 2	2 – 17
Melting Point (°C)	200 – 3500	70 – 200	2000 - 4000
Thermal Conductivity	High	Low	Medium
Thermal Expansion	Medium	High	Low
Specific Heat Capacity	Low	Medium	High
Electrical Conductivity	High	Very Low	Very Low
Tensile Strength (MPa)	100 – 2500	30 – 300	40 – 400
Tensile Modulus (GPa)	40 – 400	0.7 – 3.5	150 – 450
Hardness	Medium	Low	High
Resistance to Corrosion	Medium – Poor	Good – Medium	Good

Note: 1 Mg m^{-1} = 1000kg m^{-3}

Figure 2. 1.1 – Range of Properties

Metals:

Engineering metals are generally alloys. The term alloy is used for metallic materials formed by mixing two or more elements. For example, mild steel is an alloy of iron and carbon, stainless steel is an alloy of iron for adding elements to the iron is to improve the iron's properties. Pure metals are very weak materials. The carbon improves the strength of the iron. The presence of the chromium in the stainless steel improves the corrosion resistance.

The properties of any metal are affected by the treatment it has received and the temperature at which it is being used. Thus heat treatment, working and interaction with the environment can all change the properties. In general, metals have high electrical and thermal conductivities, can be ductile and thus permit products to be made by being bent into shape, and have a relatively high modulus of elasticity and tensile strength.

Polymers and Elastomers:

Thermoplastics soften when heated and become hard again when the heat is removed. The term implies that the material becomes plastic when heat is applied. Thermosets do not soften when heated, but char and decompose; therefore thermoplastic materials can be heated and bent to form required shapes, while thermosets cannot. Thermoplastic materials are generally flexible and relatively soft. Polythene is an example of a thermoplastic, being widely used in the forms of films or sheet for such items as bags, squeeze bottles, and wire and cable insulation. Thermosets are rigid and hard. The popular phenol formaldehyde used in the past was known as Bakelite but has largely been replaced by PP (polypropylene) and PE (polythene) in the modern industry; Bakelite is a thermoset and was widely used for electrical plug casings, door knobs and handles.

The term elastomers is used for polymers which by their structure allow considerable extensions that are reversible. The material used to make rubber bands is an obvious example of such a material.

All thermoplastics, thermosets and elastomers have low electrical conductivity and low thermal conductivity, hence their use for electrical and thermal insulation. Compared with metals, they have lower densities and higher coefficients of expansion, are generally more corrosion resistant, have a lower modulus of elasticity, tensile strengths which are nearly as high as metals, are not as hard, and give larger elastic deflections. When loaded they tend to creep, i.e. the extension gradually changes with time; their properties depend very much on the temperature so that a polymer which may be tough and flexible at room temperature may be brittle at 0°C and creep at a very high rate at 100°C.

Ceramics and Glasses:

Ceramics and glasses tend to be brittle, have a relatively high modulus of elasticity, are stronger in compression than in tension, are hard, chemically inert, and have low electrical conductivity. Glass is just a particular form of ceramic, with ceramics being crystalline and glasses non-crystalline. Examples of ceramics and glasses abound in the home in the form of cups, plates and glasses. Alumina, silicon carbide, cement and concrete are examples of ceramics; because of their hardness and abrasion resistance, ceramics are widely used for the cutting edges of tools.

Composites:

Composites are materials composed of two different materials bonded together in such a way that one serves as the matrix and surrounds the fibres or particles of the other. There are composites involving glass fibres or particles in polymers, ceramic particles in metals (referred to as cermets) and steel rods in concrete (referred to as reinforced concrete). Timber is a natural composite consisting of tubes of cellulose in a natural polymer called lignin.

Composites are able to combine the good properties of other types of materials while avoiding some of their drawbacks. Composites can be made low density, with strength and a high modulus of elasticity; however, they generally tend to be more expensive to produce.

Costs:

These can be considered in relation to the basic costs of the raw materials, the costs of manufacturing and the life and maintenance costs of the finished product.

Comparison of the basic costs of materials is often on the basis of the cost per unit weight or cost per unit volume; for example, if the cost of 10 kg of a metal is, say, $15 then the cost per kg is $1.50. f the metal has a density of 8000 kg/m^3 then 10 kg will have a volume of 10/8000 = 0.00125 m³ and so the cost per cubic metre is 1.5/0.00125 = $1200. The formulae to determine the cost per m³ can be written as:

$$\text{Cost per m}^3 = (\text{Cost per Kilogram}) \times \text{Density}$$

However, often a more important comparison is on the basis of per unit strength or cost per unit stiffness for the same volume of material; this enables the cost of, say, a beam to be considered in terms of what it will cost to have a beam of a certain strength or stiffness. Hence if, for comparison purposes, a beam of volume 1 m³ is considered then if the tensile strength of the above material is 500 MPa, the cost per MPa of strength will be 1200/500 = $2.40. The formulae to determine the same volume is:

$$\text{Cost per unit strength} = \frac{(\text{Cost/m}^3)}{\text{Strength}}$$

and similarly

$$\text{Cost per unit strength} = \frac{(\text{Cost/m}^3)}{\text{Stiffness}}$$

The costs of manufacturing will depend on the processes used. Some processes require a large capital outlay and then can be employed to produce large numbers of the product at a relatively low cost per item. Others may have little in the way of setting-up costs but a large cost per unit product. The cost of maintaining a material during its life can often be a significant factor in the selection of materials. A feature common to many metals is the need for a surface coating to protect them from corrosion by the atmosphere. The rusting of steels is an obvious example of this and dictates the need for such activities as the continuous repainting of the Sydney Harbour Bridge.

Figure 1.14

In Figure 1.14 above, the main supporting arch is showing clear signs of rust and needs immediate descaling and repainting.

Example

On the basis of the following data, compare the costs per unit strength of the two materials for the same volume of material.

    Low-carbon Steel: Cost per kg $1.00, density 7800 kg/m³, strength 1000 MPa

    Aluminium Alloy (Mn):   Cost per kg $2.20, density 2700 kg/m³, strength 200MPa

For the steel, the volume of 1 kg is 1/7800 = 0.00013 m³ and so the cost per m³ is 1/0.00013 = $7692. The cost per MPa of strength is thus 7692/1000 = $7.69. For the aluminium alloy, the volume of 1 kg is 1/2700 = 0.00037 m³ and so the cost per m³ is 02.2/0.00037 = $5946; therefore, although the cost per kg is greater than that of the *steel, because of the lower density the cost per cubic metre is less. The cost per MPa of strength is 5946/200 = $29.73. On a comparison on the strengths of equal volumes, it is cheaper to use the steel where $7.69 < $29.73.*

MEM30007A - Select common engineering materials
Topic 1 - Properties of Materials

Review Problems:

MEM30007-RQ-01

1. What types of properties would be required for the following products?
   (a) A domestic kitchen sink.
   (b) A shelf on a bookcase.
   (c) A cup.
   (d) An electrical cable.
   (e) A coin.
   (f) A car axle.
   (g) The casing of a telephone.

2. For each of the products listed in problem 1, identify a material that is commonly used and explain why its properties justify its choice for that purpose.

3. Which properties of a material would you need to consider if you required materials which were:
   (a) Stiff,
   (b) Capable of being bent into a fixed shape.
   (c) Capable of not fracturing when small cracks are present.
   (d) Not easily broken.
   (e) Acting as an electrical insulator.
   (f) A good conductor of heat.
   (g) Capable of being used as the lining for a tank storing acid.

4. A colleague informs you that a material has a high tensile strength with a low percentage elongation. Explain how you would expect the material to behave.

5. A colleague informs you that a material has a high tensile modulus of elasticity and good fracture toughness. Explain how you would expect the material to behave.

6. What is the tensile stress acting on a strip of material of cross-sectional area 50 mm^2 when subject to tensile forces of 1000 N?

7. Tensile forces act on a rod of length 300 mm and cause it to extend by 2 mm. What is the strain?

8. An aluminium alloy has a tensile strength of 200 MPa. What force is needed to break a bar of this material with a cross-sectional area of 250mm^2?

9. A test piece of a material is measured as having a length of 100 mm before any forces are applied to it. After being subject to tensile forces it breaks and the broken pieces are found to have a combined length of 112 mm. What is the percentage elongation?

10. A material has a yield stress of 250 MPa. What tensile forces will be needed to cause yielding if the material has a cross-sectional area of 200 mm^2?

11. A sample of high tensile brass is quoted as having a tensile strength of 480 MPa and an elongation of 20%. An aluminium-bronze is quoted as having a tensile strength of 600 MPa and an elongation of 25%. Explain the significance of these data in relation to the mechanical behaviour of the materials.

12. A grey cast iron is quoted as having a tensile strength of 150 MPa, a compressive strength of 600 MPa and an elongation of 0.6%. Explain the significance of the data in relation to the mechanical behaviour of the material.

13. A sample of carbon steel is found to have an impact energy of 120 J at temperatures above 0°C and 5 J below it. What is the significance of these data?

14. Mild steel is quoted as having an electrical resistivity of 1.6×10^{-7}. Is it a good conductor of electricity?

15. A colleague states that he needs a material with a high electrical conductivity. Electrical resistivity tables for materials are available. What types of resistivity values would you suggest he looks for?

16. Aluminium has a resistivity of $2.5 \times 10^{-8}\,\Omega\,m$. What will be the resistance of an aluminium wire with a length of 1 m and a cross-sectional area of 2 mm²?

17. How do the properties of thermoplastics differ from those of thermosets?

18. You read in a textbook that "Designing with ceramics presents problems that do not occur with metals because of the almost complete absence of ductility with ceramics". Explain the significance of the comment in relation to the exposure of ceramics to forces.

19. Compare the specific strengths, and costs per unit strength for equal volumes, for the materials giving the following data:
Low-carbon steel: Cost per kg $0.10, density 7800 kg/m³, strength 1000 MPa
Polypropylene: Cost per kg $0.20, density 900 kg/m³, strength 30 MPa.

Topic 2 – Properties Data:

Required Skills:
- Recognize the role played by standards issued by national and international standards bodies in aiding effective communication between material producers, product manufacturers and consumers.
- Identify the range of sources from which information about the properties of materials can be obtained.
- In considering the suitability of a material for a particular application, determine what information is required and gather, collate, analyse and present it.

Required Knowledge:
- Use tables, charts and specifications.

Standards:
There are many thousands of standards laid down by national standards bodies such the Standards Australia (AS) and international bodies such as the International Organization for Standardization (ISO). A standard is a technical specification drawn up with the cooperation and general approval of all interests affected by it with the aims of benefiting all concerned by ensuring consistency in quality, rationalizing processes and methods of operation, promoting economic production methods, providing a means of communication, protecting consumer interest, encouraging safe practices, and helping confidence in manufacturers and users. For example, the Australian Standard AS1210-2010 is a standard setting out the minimum requirements for the materials, design, manufacture, testing, inspection, certification, documentation and dispatch of fired and unfired pressure vessels constructed in ferrous or non-ferrous metals by welding, brazing, casting, forging, or cladding and lining and includes the application of non-integral fittings required for safe and proper functioning of pressure vessels.

Standards, both national and international, are used in the case materials to ensure such things as consistency of quality and in the use of terms and rationalization of testing methods, and provide an efficient means of communication between interested parties; therefore, if a material is stated by its producer to be to a certain standard, tested by the methods laid down by certain standards, then a customer need not have all the written details of what properties and tests have been carried out by the producer in order to know what properties to expect of the material. The Australian Standard for tensile testing of metals is AS1391-1991 and lays down such items as the sizes of the test pieces to be used (refer to Topic 3 – Materials Testing:). There are standards for materials such as copper and copper alloy plate (AS 1566-1997), steel plate, sheet and strip (AS/NZS 3678-2011), the plastic polypropylene (AS/NZS 5065-2005), etc. which lay down the composition and properties for such materials.

Data Sources:
Data on the properties of materials is available from a range of sources. These include:

1. Specifications issued by bodies responsible for standards, e.g. Australian Standards Association, the British Standards Institution, the American Society for Metals, the International Organization for Standardization, etc. The standards operating in Australia are those issued by the Standards Australia listing all the

standards and also whether they are Australian, American, European or international standards.

2. Data books, e.g:
   - *The Mechanical Properties of Metals* by C.W. Lung and N.H. March, (World Science Publishing, 1999).
   - *The Theory of the Properties of Metals and Alloys* by N.F. Mott and H. Jones, (General Publishing Company, 1986), *Handbook of Plastics and Elastomers* edited by C. A. Harper (McGraw-Hill, 1975.,
   - *Smithells Metals Reference Book* by W.F. Gale and T.C. Totemier (Elsevier Butterworth-Heinemann, 2004).
   - *Hydrogen in Metals* by H. Wipf, (Springer-Verlag, 1997).

3. Computerized databases which give materials and their properties with means to rapidly access particular materials or find materials with particular properties, e.g. *Cambridge Materials Selector* (Cambridge University Engineering Department, 1992), *MAT.DB* (American Society for Metals, 1990).
   - *One Steel Catalogue*, www.onesteel.com/home.asp
   - *Alcoa Products*, www.alcoa.com
   - *Plastics Database*, www.plasticsintl.com/sortable_materials.php?display=thermal

4. Trade associations,
   - Australian Mines and Metals Association
   - Australian Plastics and Rubber Institute
   - Standards Australia
   - Australian Construction Industry
   - Copper Development Association

5. Data sheets supplied by suppliers of materials.

6. In-company tests - used to check samples of a bought-in material to ensure that it conforms to the standards specified by the supplier.

Coding Systems:

Coding systems are used to refer to particular metals. Such codes can relate to die chemical composition or the type of properties it has and are a concise way of specifying a particular material without having to write out its full chemical composition or properties.

Steel:

The standard for Structural Steel for general engineering purposes is AS 3678-2011.

Steels are referred to in terms of a code specified by the Australian Standards:

Mechanically Tested Grades:

Mechanically tested carbon and carbon-manganese grades and low alloy (weathering) grades are designated AS/NZS 3678-250 or AS/NZS 3678-WR350 where:
- AS/NZS 3678 denotes the standard number.
- 250, 350, 410 denotes the grade.
- WR denotes weather resistant.

Additional Properties:

The grade designation for the steel may also indicate mechanical testing properties:

Through-thickness tensile properties: e.g. AS/NZS 3678-250Z25 where:
- Z indicates the material has been thickness tested.
- XX indicated the value of the minimum reduction in area.

Impact properties: e.g. AS/NZS 3678-250L0 and AS/NZS 3678-350Y40 where:
- L or Y indicates the material has been impact tested.

- XX indicates the value of a temperature equal to the actual impact test temperature that is at or below 0°C.

Analysis Grades:

For analysis steels is a five-digit alpha numeric system; e.g. AS/NZS 3678-A1006.

The first character is a letter denoting deoxidization practice:
- A denotes aluminium killed.
- K denotes silicon killed, with or without aluminium additions

The first two digits numbers in the series indicates the type of steel:
- 10 indicates plain carbon steel.
- 15 indicates carbon-manganese steel.

The last two digits indicate the approximate mean of the specified carbon range based on the corresponding AISE-SAE J403 grade.

Modification Symbol:

The modification symbol "X" may be added to the grade designation to indicate a major deviation in chemical composition of any grade; e.g. AS/NZS 3678-XK1016.

Fine Grained Steel:

The letters "FG" added to the grade designation indicates the steel was produced using fine grained steel making practices; e.g. AS/NZS 3678-250FG or AN/NZS 3678-K1016FG.

Stainless Steel:

Stainless Steel is available in five groups; Austenitic, Ferritic, Martensitic, Duplex (Austenitic-Ferritic) and Precipitation Hardening. Available grades are 304, 304L, 310, 316, 316L, 409, 1.4003, 430 and S31803.

Aluminium:

The standard for wrought aluminium and aluminium alloy flat sheet, coiled sheet and plate for general engineering purposes is AS 1734-1997.

The alloy designation system consists four digits, e.g. AS 1734/4043:
- The first digit indicates the alloy group in accordance with the following system:
    - 1XXX – Aluminium, 99.0% min.
    - 2XXX – Aluminium alloy – copper.
    - 3XXX – Aluminium alloy – copper and silicon.
    - 4XXX – Aluminium alloy – silicon.
    - 5XXX – Aluminium alloy – magnesium.
    - 6XXX – Aluminium alloy – magnesium and silicon.
    - 7XXX – Aluminium alloy – zinc.
    - 8XXX – Aluminium alloy – other alloying elements.
- The second digit denotes the impurities. If the digit is zero, it indicates unalloyed aluminium having natural impurity limits with no special control exercised on individual impurities. Integers 1 to 9 inclusive, which are assigned consecutively as needed, indicate special control of one or more individual impurities.
- The third and fourth digits in the designation indicate the minimum aluminium percentage.

The temper designation is a letter and digit system denoting indicating the sequence of basic treatments used to produce the tempers; e.g. AS 1734/4043-H21:

- The first letter indicates the basic temper:
    - F – as fabricated
    - O – annealed
    - H – strain hardened

T – thermally treated
- The first number after H indicates the specific combination of basic operations:
  - H1 – strain hardened only.
  - H2 – strain hardened and then partially annealed.
  - H3 – strain hardened and then stabilized.
  - H4 – strain hardened and then lacquered or painted.
- The digit following the designations H1, H2 and H3 indicates either annealed or the final degree of strain hardening, as follows:
  - 0 – annealed tempers.
  - 8 – tempers having a final degree of strain-hardening equivalent to that resulting from approximately 75% reduction of area. Also known as 'fully hard'.
  - 9 – extra hard tempers.
- The third digit, when used, indicates a variation of a two-digit H temper, and is added when the degree of control of temper, or the mechanical properties vary slightly from those corresponding to the two-digit temper designation:
  - H111 – strain-hardened less than the amount required for a controlled H11 temper.
  - H112 – manufactured without special control over the amount of strain-hardening or thermal treatment but when products acquire some temper from the shaping processes. Products manufactured to this temper are subject to mechanical testing and are required to meet mechanical property limits.
  - H116 – having characteristics that substantially eliminate the susceptibility of the alloy to exfoliation under marine corrosion conditions.
  - H121 – strain hardened less than the amount required for a controlled H12 temper.
  - H311 – strain hardened less than the amount required for a controlled H31 temper.

Copper:

Copper is used for tubing, fittings and wrought products.

Tubes:

Tubes are designated by the standard number, followed by the nominal size and then the classification.

Australian Standard – AS 1432.

Nominal Size – DN followed by 6, 8, 10, 15, 18, 20, 25, 32, 40, 50, 65, 80, 90, 100, 125, 15 or 200.

Classification types – A, B, C, or D.

Wrought Products:

The Australian Standard for wrought products is AS 2738-2000. For copper refinery alloys, the designations comprise the letters C followed by a hyphen and additional characters which indicate the refining process. For wrought and cast coppers and copper alloys, the designations are in accordance with the Unified Numbering System.

    Wrought Alloys
- Coppers – C10000 to C15900
- High Copper alloy – C16000 to C 19900
- Brasses – C20000 to C49900
- Bronze – C50000 to C66300
- Miscellaneous copper-zinc alloy – C66400 to C69900

Copper-nickel alloy – C70000 to 72900
Copper-nickel-zinc alloy – C73000 to C79900
Cast Alloys
Coppers and copper alloys – C80000 to C82990
Brass – C83000 to C87900
Copper-Tin alloy – C90000 to C91900
Copper-Tin-Lead alloy – C92000 to C94500
Aluminium Bronze – C95000 to C95900
Copper-Nickel alloy – C96000 to C96950

Plastics:

Polyamide Pipes:

Polyamide pipes are used in gas mains and services for direct burial and reliner applications; the pipes are intended for use in the distribution of natural gas, manufactured gas, liquefied petroleum gas (LPG) and LPG/air mixtures at pressures up to 400 kPa.

Classification:
Class 300 for pressures up to 300kPa
Class 400 for pressures up to 400kPa

Timber:

All species of timber used in Australia is sourced locally and imported and include "Softwood" which is classified botanically as *Gymnospermae* and "Hardwood" which is classified botanically as *Angiospermae*.

In plans, specifications and official documents, timber should be specified by the botanical name rather than the common name; common timbers being used in Australia:

Acacia Baileyanna	Cootamundra Wattle	Eucalyptus regnans	mountain ash
Acer	Maple	Eucalyptus resinifera	red mahogany
Agathis Australis	New Zealand Kauri	Eucalyptus saligna	Sydney blue gum
Callitris glaucophylla	cypress pine,	Eucalyptus sideroxylon	red ironbark
Eucalyptus camaldulensis	red river gum	Guaiacum	lignum vitae
Eucalyptus dalrympleana	mountain gum	Leptospermum laevigatum	coast tea-tree
Eucalyptus delegatensis	alpine ash	Nothofagus solandri	mountain beech
Eucalyptus diversicolor	karri	Pinus radiata	radiata pine
Eucalyptus globulus	southern blue gum	Pseudotsuga menziesii	Douglas Fir (Oregon)
Eucalyptus largiflorens	black box	Tectona grandis	Teak
Eucalyptus macrorhyncha	Red Stringybark	Syncarpia glomulifera	turpentine
Eucalyptus marginata	jarrah	Thuja plicata	western red cedar
Eucalyptus microcorys	tallowwood		

Alloy Steel:

Alloy steel is steel that is combined with a variety of elements in total amounts between 1.0% and 50% by weight to improve the mechanical properties of the steel. Alloy steels are broken down into two groups: low-alloy steels and high-alloy steels. The difference between the two is somewhat arbitrary; one group of scientists will define the difference at 4.0%, while another group will define it at 8.0%. Most commonly, the phrase "alloy steel" refers to low-alloy steels.

All steels are truly alloys, but not all steels are called "alloy steels"; even the simplest steels are iron (Fe) (about 99%) alloyed with carbon (C) (about 0.1% to 1%, depending on type). However, the term "alloy steel" is the standard term referring to steels with *other* alloying elements *in addition to* the carbon. Common alloy elements include manganese (the most common one), nickel, chromium, molybdenum, vanadium, silicon, and boron. Less common alloy elements include aluminum, cobalt, copper, cerium, niobium, titanium, tungsten, tin, zinc, lead, and zirconium.

The following is a range of improved properties in alloy steels (as compared to carbon steels): strength, hardness, toughness, wear resistance, corrosion resistance hardenability, and hot hardness. To achieve some of these improved properties the metal may require heat treating.

Some of these find uses in exotic and highly-demanding applications, such as in the turbine blades of jet engines, in spacecraft, and in nuclear reactors. Because of the ferromagnetic properties of iron, some steel alloys find important applications where their responses to magnetism are very important, including in electric motors and in transformers.

Alloying elements also have an effect on the eutectoid temperature of the steel. Manganese and nickel lower the eutectoid temperature and are known as *austenite stabilizing elements*. With enough of these elements the austenitic structure may be obtained at room temperature. Carbide-forming elements raise the eutectoid temperature; these elements are known as *ferrite stabilizing elements*.

The following table indicates the principal effects of major alloying elements for steel.

Element	Percentage	Primary Function
Aluminium	0.95 – 1.30	Alloying element in nitriding steels.
Bismuth	-	Improves machinability.
Boron	0.001 – 0.003	A powerful hardenability agent.
Chromium	0.5 – 2	Increases hardenability
	4 – 18	Increases corrosion resistance
Copper	0.1 – 0.4	Corrosion resistance
Lead	-	Improved machinability
Manganese	0.25 – 0.40	Combines with sulphur and with phosphorus to reduce the brittleness. Also helps to remove excess oxygen from molten steel.
	> 1	Increases hardenability by lowering transformation points and causing transformations to be sluggish.
Molybdenum	0.2 - 5	Stable carbides; inhibits grain growth. Increases the toughness of steel, thus making molybdenum a very valuable alloy metal for making the cutting parts of machine tools and also the turbine blades of turbojet engines. Also used in rocket motors.
Nickel	2 – 5	Tougher.
	12 – 20	Increases corrosion resistance.
Silicon	0.2 – 0.7	Increases strength
	2.0	Spring Steels.
	Higher Percentages	Improves magnetic properties

Sulphur	0.08 – 0.15	Free-machining properties.
Titanium	-	Fixes carbon in inert particles; reduces martensitic hardness in chromium steels
Tungsten	-	Also increases the melting point.
Vanadium	0.15	Stable carbides; increases strength while retaining ductility; promotes fine grain structure. Increases the toughness at high temperatures.

Mechanical Property Requirements for Steels Heat-Treated then Cold Finished

Alloy Designation AS 1444	Mechanical Property Designation	Limiting Ruling Section	Tensile Strength MPa		0.20% Proof Stress MPa	Elongation %	Brinell Hardness	
		mm	min	max	min	min	min	max
4140	R	63	700	850	525	12	201	255
	S	63	770	930	585	11	223	277
	T	63	850	1000	680	9	248	302
	U	63	930	1080	755	9	269	331
	V	63	1000	1150	850	9	293	352
4340	T	63	850	1000	680	9	248	302
	U	63	930	1080	755	9	269	331
	V	63	1000	1150	850	9	293	352

Mechanical Property Requirements for Steels in the Heat-Treated Condition

Alloy Designation AS 1444	Mechanical Property Designation	Limiting Ruling Section	Tensile Strength MPa		0.20% Proof Stress MPa	Elongation %	Izod Impact	Charpy Impact	Brinell Hardness	
		mm	min	max	min	min	min	min	min	max
X4036	R	250	700	850	480	15	34	28	201	255
	R	150	700	850	510	17	54	50	201	255
	S	100	770	930	570	15	54	50	223	277
	T	63	850	1000	665	13	54	50	248	302
	U	30	930	1080	740	12	47	42	269	331
	V	20	1000	1150	835	12	47	42	293	352
4130	R	150	700	850	510	17	54	50	201	255
	S	100	770	930	570	15	54	50	223	277
	T	63	850	1000	665	13	54	50	248	302
	U	30	930	1080	740	12	47	42	269	331
4140	R	250	700	850	480	15	34	28	201	255
	S	250	700	850	540	13	27	22	233	277
	S	150	770	930	570	15	54	50	233	277
	T	100	850	1000	665	13	54	50	248	302
	U	63	930	1080	740	12	47	42	269	331
	V	30	1000	1150	835	12	47	42	293	352
	W	20	1080	1230	925	12	40	35	311	375
4340	T	250	850	1000	635	13	40	35	248	302
	T	150	850	1000	665	13	54	50	248	302
	U	63	930	1080	740	12	47	42	269	311
	V	30	1000	1150	835	12	47	42	293	352
	W	30	1080	1230	925	11	41	38	311	375
	X	30	1150	1300	1005	10	34	28	341	401
	Y	30	1230	1380	1080	10	24	20	363	429
	Z	30	1550	-	1125	5	10	9	444	-

Topic 2 - Properties Data

Alloy Designation AS 1444	Mechanical Property Designation	Limiting Ruling Section	Tensile Strength MPa		0.20% Proof Stress MPa	Elongation %	Izod Impact	Charpy Impact	Brinell Hardness	
		mm	min	max	min	min	min	min	min	max
X7039	R	150	700	850	510	17	54	-	201	255
	S	100	770	930	570	15	54	-	223	277
	T	63	850	1000	655	13	47	-	248	302
X7232	U	250	930	1080	740	12	40	35	269	331
	U	150	930	1080	740	12	47	42	267	331
	V	150	1000	1150	835	12	47	42	293	352
	W	100	1080	1230	925	11	40	35	311	375
X9931	T	250	850	1000	635	13	40	35	248	302
	T	150	850	1000	665	13	54	50	248	302
	U	250	930	1080	725	12	34	28	269	331
	U	150	930	1080	740	12	47	42	269	331
	V	150	1000	1150	835	12	47	42	293	352
	W	63	1150	1230	925	11	40	35	311	375
	X	150	1150	1300	1005	10	34	28	341	401
	Y	150	1230	1380	1080	10	34	28	363	429
	Z	100	1550	-	1125	7	13	11	444	-

Data Analysis:

The procedure that might be adopted in searching for a material with the properties required for a particular product could be:
1. Identify the properties required.
2. Look in Australian or international standards, or data books or computer databases for materials with the required properties.
3. The above might refine the search to a particular material; if further information is required, trade association publications or supplier data sheets can be consulted.

To illustrate the above, consider the search for a material for use as a conductor of electricity where high conductivity is the main property required. Since metals are, in general, good conductors and polymers and ceramics very poor ones then the choice would seem to be among metals. When tables are consulted the following information can be found for electrical conductivities, at 20°C:

Aluminium	40×10^6 S/m
Copper	64×10^6 S/m
Gold	50×10^6 S/m
Iron	11×10^6 S/m
Silver	67×10^6 S/m

If cost is also a factor then gold and silver are likely to be ruled out leaving copper appearing to be the optimum choice. If we now consider what form of copper then tables are likely to yield data in the following form:

C101	Electrolytic tough-pitch h.c. copper	101.5 - 100
C103	Oxygen-free h.c. copper	101.5 - 100
C105	Phosphorus deoxidized arsenical copper	50 - 35
C108	Copper-cadmium	92 - 80

The conductivities are not expressed in S/m but in units called IACS (international annealed copper standard) units and written as a percentage; this scale is based on 100% being the conductivity (or resistivity) of annealed copper at 20°C and is a resistivity of 1.7241×10^8 W m or a conductivity of 58.00×10^6 S/m. Therefore, if there are no other considerations C101 or C103 would appear to be the choice. Often, however, there are other factors to be taken into account, such as strength.

Consider another example, a requirement for a metal with a low melting point. The aim is to use the metals in die casting for the production of small components for toys, e.g. toy car steering wheels and drive shafts. The following are some of the melting points for metals that can be found from tables:

Aluminium	600°C
Lead	320°C
Magnesium	520°C
Zinc	380°C

Lead and zinc have the lowest melting points. If we add another requirement of reasonable strength in the as-cast condition then tables give:

Lead	Tensile strength 20 MPa
Zinc	Tensile strength 280 MPa

Thus, taking strength into account, zinc would appear to be the choice. There are, however, other considerations that have to be taken into account before a choice can be made and will be covered in Topic 6 – Selection of Materials:

Material	Tensile Strength (MPa)	Yield Stress (MPa)	% Elongation
Aluminium	110	95	14
Brass	250	115	30

Copper		220	70	15
Mild Steel	GR250	250	320	13
	GR350	350	430	12
	GR450	450	500	8
Nylon		75	45	8
Stainless Steel	304	650	270	58
	316	605	260	55
	430	515	325	28

Review Problems:

MEM30007-RQ-02

1. The rainwater guttering used for buildings is required to have a high stiffness per unit weight so that it does not sag under its own weight. Use tables to obtain the specific modulus or values of the modulus and density and hence compare cast iron, aluminium alloys and the plastic PVC as possible materials.

2. A car bodywork panels for need to be in sheet form and stiff. Use tables to obtain modulus of elasticity values and then compare carbon steel, an aluminium alloy, polypropylene and a composite formed by polyester with 65% glass fibre cloth.

3. The fan in a vacuum cleaner needs to be made of a low-density material and a high tensile strength, i.e. a high specific strength. The aluminium alloy LM6 has been suggested because the fan could then be die cast. Use to the tables to obtain the specific strength of the material.

4. The plastic ABS has been suggested for use as the casing for a radio. The properties required include high stiffness. Determine from tables the modulus of elasticity and compare it with other plastics.

5. The material high-tensile brass HTB1 has been suggested for use as a marine propeller. Use tables to obtain values of its tensile strength, 0.1% proof stress and percentage elongation.

6. Determine the percentage and element to alloy to steel to create spring steel.

7. Select the designation for aluminium-manganese unalloyed aluminium having natural impurity limits with no special control exercised on individual impurities containing minimum 52% aluminium.

8. Using the following table which lists the properties for polymers; select from the list a material which will be stiff and not too brittle.

Polymer	Tensile Strength (MPa)	Tensile Modulus (MPa)	Elongation %
ABS	17 – 58	1.4 – 3.1	10 – 140
Acrylic	50 – 70	2.7 – 3.5	5 – 8
Cellulose Acetate	26 – 65	1.0 – 2.0	56 – 55
Cellulose Acetate Butyrate	18 – 48	0.5 – 1.4	40 – 90
Polyacetal Homopolymer	70	3.6	15 – 75
Polyamide, Nylon 6	75	1.1 – 3.1	60 - 320
Polyamide, Nylon 66	80	2.8 – 3.3	60 - 300

9. Using the following table which lists the properties for steels which have been quenched and tempered to 550-660°C; select, from the list the strongest steel.

Steel	Tensile Strength (MPa)	Yield Modulus (MPa)	Elongation %
212M36	550 – 700	340	20
216M36	550 – 700	380	15
226M44	700 – 850	450	15

Topic 3 – Materials Testing:

Required Skills:
- Identify the standard tests used for key properties and materials.
- Recognize the test procedures required.
- Interpret test results explaining their validity and limitations

Required Knowledge:
- Reading graphs, tables and charts.
- Testing procedures.
- Processing formulae.

Standard Tests:
The standard tests used in Australia are those specified by the Australian Standards Association; these include:

AS1146-1990	Methods for impact tests on plastics
AS1391-1911	Methods for tensile testing of metals
AS1544-2003	Method for impact tests on metals
AS1815-2007	Metallic materials – Rockwell hardness test
AS1816-2007	Metallic materials – Brinell hardness test
AS1817-2003	Metallic materials – Vickers hardness test
AS2505-2004	Metallic materials – Sheet, strip and plate – Bend tests
ASTM E1820	Standard test method for measurement of fracture toughness
ASTM D638-10	Standard test method for tensile properties of plastics
ISO 12135	Metallic materials. Unified method for the determination of quasi-static fracture toughness

The Tensile Test:
In a tensile test, measurements are made of the force required to extend a standard test piece at a constant rate, the elongation of a specified gauge length of the test piece being measured by some form of extensometer. Australian Standards state that the rate at which the stresses are applied should be between 2 and 10 MPa/s if the tensile modulus is less than 150 GPa and between 6 and 30 MPa/s if the tensile modulus is equal to or greater than 150 GPa. In order to eliminate any variations in tensile test data due to differences in the shapes of test pieces, standard shapes and sizes are adopted.

The Test Piece:
Test pieces are said to be *proportional* if the relationship between the gauge length L_o and the cross-sectional area A of the gauge length:

$$L_o = kSA$$

Australian standards specify that the constant k should have the value 5.65 and the gauge length should be 20 mm or greater. With circular cross-sections $A = \frac{1}{4}\pi d^2$ and thus to a reasonable approximation this value of k gives:

$L_o = 5d$

With circular cross-sectional areas which are too small for this value of *k,* a higher value may be used, preferably 11.3. When test pieces are proportional the same test results are given for the same test material when different size test pieces are used.

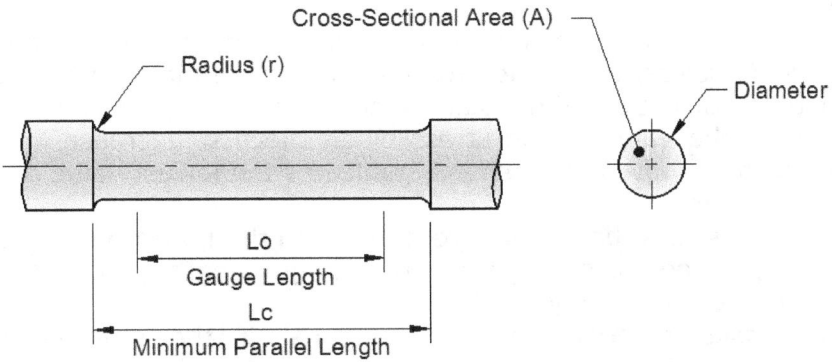

Figure 3.1 - Round Test Piece

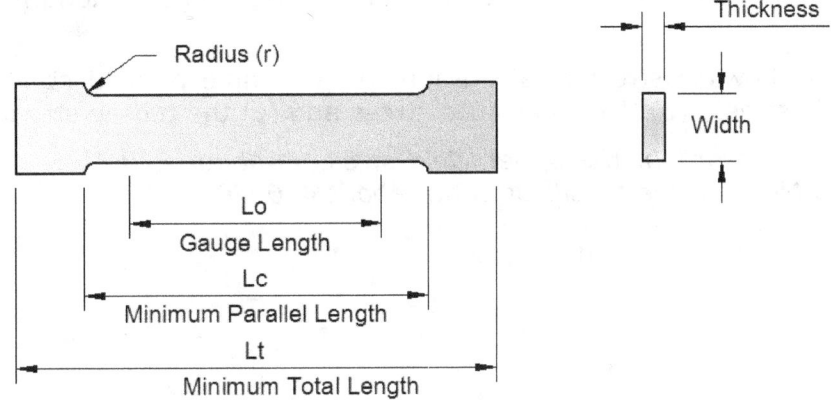

Figure 3.2 – Flat Test Piece

Figure 3.1 and Figure 3.2 show the standard size test pieces for round and flat samples of metals while Table 3.1 lists the standard dimensions that can be used. For the tensile test data for the same material to give essentially the same results, regardless of the length of the test piece used, it is vital that the standard dimensions are adhered to. An important feature of the dimensions is the radius given for the shoulders of the test pieces. Very small radii can cause localized stress concentrations which may result in the test piece failing prematurely.

Flat Test Pieces

b mm	Lo mm	Lc mm	Lt mm
20	80	120	140
12.5	50	75	87.5

Round Test Pieces

d mm	A mm²	Lo mm	Lc mm
20	314.2	100	110
10	78.5	50	55
5	19.6	25	28

Figure 2. 2 – Dimensions of Standard Test Pieces

Tensile Test Results:

The results of tensile tests are obtained as force-extension data which are generally plotted, either manually or by a machine, as a force-extension graph. Since stress is force/original area and strain is extension/original gauge length then the graph is readily translated into stress-strain. From such graphs the following quantities can be determined:

1. The *tensile strength*, being the stress corresponding to the maximum force.
2. The *yield stress*, being the stress at which the material begins to yield and show plastic deformation without any increase in load. The term *upper yield stress* is used for the value of the stress when the first decrease in force at the yield is observed and *lower yield stress* implicates the lowest value of stress during plastic yielding.
3. The *proof stress*, being the stress at which the non-proportional extension is equal to a specified percentage of the gauge length. 0.1 % and 0.2% are the percentages commonly used.
4. The *tensile modulus*, being the slope of the stress-strain graph over its proportional region.

In addition, measurements of the gauge length before and after breaking enable the *percentage elongation* to be determined, this being the permanent elongation of the gauge length expressed as a percentage of the original gauge length.

Example

Figure 3.3 shows a stress-strain graph for a sample of mild steel. Determine (a) the upper yield sttess; (b) the lower yield stress and (c) the tensile strength.

Reading off the graph, the upper yield stress is about 270 MPa, the lower yield stress about 220 MPa and the tensile strength about 420 MPa.

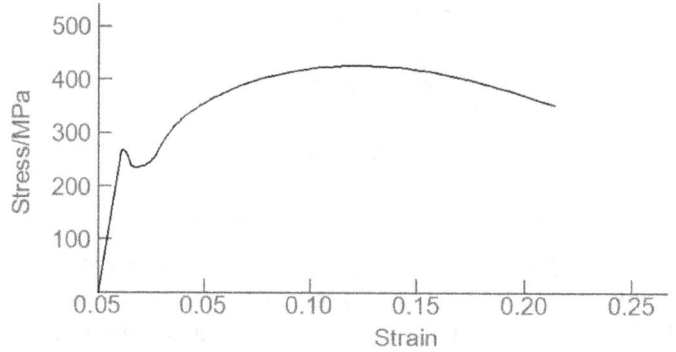

Figure 3.3 – Stress-Strain for a Mild Steel Sample

Example

Figure 3.4 shows part of the stress-strain graph for a sample of an aluminium alloy. Determine the 0.1 % and 0.2% proof stresses.

The 0.1 % proof stress is about 460 MPa and the 0.2% about 520 MPa.

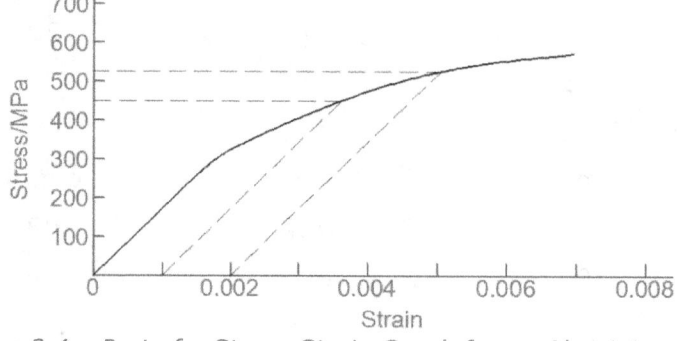

Figure 3.4 – Part of a Stress-Strain Graph for an Aluminium Alloy

Validity of Tensile Test Data:

The purpose of taking tensile test pieces and carrying out the tests is to obtain data which enable judgements to be made about the material from which the test piece was cut. The samples of a material have to be taken in such a way that the properties deduced from the tensile test are representative of the material as a whole; assuming this however may cause some problems. The following paragraphs outline some of these problems.

The properties of a product may not be the same throughout the part as in a casting there may be different cooling rates in different areas of the casting, e.g. the surface compared with the core, or thin sections compared with thick sections. As a result, the internal structure of the material may differ and, as a consequence, the tensile properties differ. A tensile test piece cut from one part may not thus represent the properties of the entire casting. For the same reason, the properties of a separately cast test piece may not be the same as those of the cast product because the different sizes of the two lead to different cooling rates.

Note that if the mechanical properties of metals are looked up in tables different values of the properties may be quoted for different limiting ruling sections. The *limiting ruling section* is the maximum diameter of a round bar at the centre of which the specified properties may be obtained. The reason for the difference of mechanical properties of the same material for bars of different diameters is that during the heat treatment different rates of cooling occur at the centres of such bars due to their variations in sizes. Consequently there are differences in microstructure and hence in mechanical properties. For example, the steel 070M.55 with a limiting ruling section of 19 mm may have tensile strengths of 850 to 1000 MPa, with 63 mm the strengths are 777 to 930 MPa and with 100 mm they are 700 to 830 MPa.

The properties of a product may not be the same in all directions; therefore, for example, with rolled sheet there is a directionality of properties with the tensile properties in the longitudinal, transverse and through-thickness directions of the sheet differing. With rolled brass strip we might have tensile strengths of 740 MPa in the direction of the rolling and 850 MPa at right angles to it.

The temperature in service of the product may not be the same as that of the test piece when the tensile test data was obtained. The tensile properties of metals depend on temperature. In general, the tensile modulus and tensile strength both decrease with an increase in temperature and the percentage elongation tends to increase.

The rate of loading of a product may differ from that used with the test piece. The data obtained from a tensile test are affected by the rate at which the test piece is stretched, so in order to give a standardized result the tests are carried out at a constant stress rate, between 2 and 20 MPa/s if the tensile modulus is less than 150 GPa and between 6 and 30 MPa/s if I tis equal to or greater than 150 GPa.

Interpreting Tensile Test Data:

The results from tensile tests can be used to determine the safe stresses to which a material can be subject; the higher the yield stress of a metal, the higher the stresses that it can be exposed to in service without yielding. Another important deduction that can be made is whether the material is brittle or ductile (refer to Topic 1 – Properties of Materials: Figure 1.10 – Brittle Materials and Figure 1.11 – Ductile Materials). A brittle material will show little plastic behaviour and have a low percentage elongation. A ductile material will show considerable plastic behaviour and have a high percentage elongation.

A consequence of the heat treatment and working of a material that occurs during the fabrication of products is a change in mechanical properties. Tensile test data enables the effectiveness of heat treatments and the effects of working to be monitored.

Example

Which of the materials shown in Figure 3.5 is (a) the most ductile; (b) the most brittle; (c) the strongest; (d) the stiffest?

Answers
(a) Material C, because it has the greatest plastic region to its graph,
(b) Material B, because it has the least plastic region,
(c) Material A, because it can experience the highest stress,
(d) Material A, because it has the steepest slope for its proportional region.

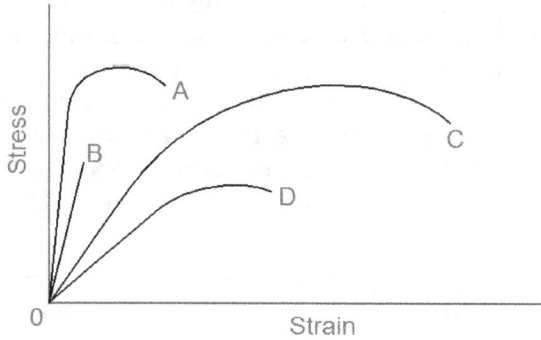

Figure 3.5

Tensile Tests for Plastics:

Tensile tests can be used with plastic test pieces to obtain stress-strain data. The term *tensile strength* has the same meaning as with metals. However, the tensile modulus, i.e. the slope of the stress-strain graph over the proportional region, cannot always be easily obtained. For many plastics there is no really straight-line part of the stress-strain graph; therefore, as a measure of the stiffness of the material, a modulus is defined in a different way. The *secant modulus* is obtained by dividing the stress at a strain of 0.2% by that strain, as illustrated in Figure 3.6.

The stress-strain properties of plastics are much more dependent than metals on the rate at which the strain is applied. For example, the tensile test may indicate a yield stress of 62 MPa when the rate of elongation is 12.5 mm/min but 74 MPa when it is 50 mm/min; also, the form of the stress-strain graph may change with a ductile material at low strain rates, becoming a brittle one at high strain rates. Figure 3.7 and Figure 3.8 show the general forms of stress-strain graphs for plastics at different strain rates; Figure 3.7 shows the effect of a brittle plastic while Figure 3.8 shows a ductile plastic. Another factor that is more marked than with metals is the effect of temperature on the properties of plastics.

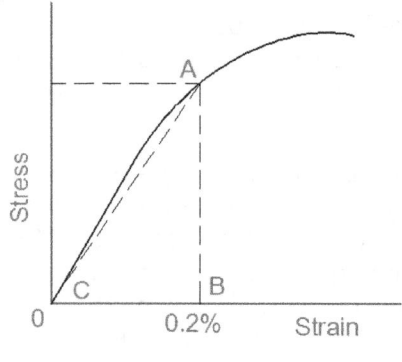
Figure 3.6

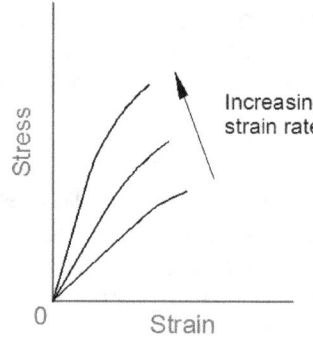
Figure 3.7

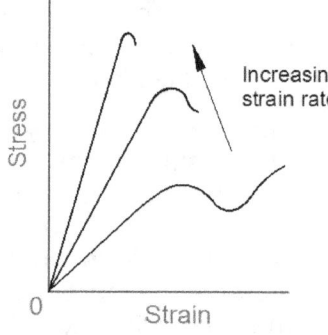
Figure 3.8

Topic 3 - Materials Testing

Example

Figure 3.9 is the stress-strain graph for a sample of Novodur by Bayer (UK) Ltd grade ABS resin used mainly for housings and covers requiring good toughness, strength, stiffness, chemical resistance and a good to very good surface finish.   (courtesy of) .
Estimate:
  (a) the tensile modulus and
  (b) the tensile strength.

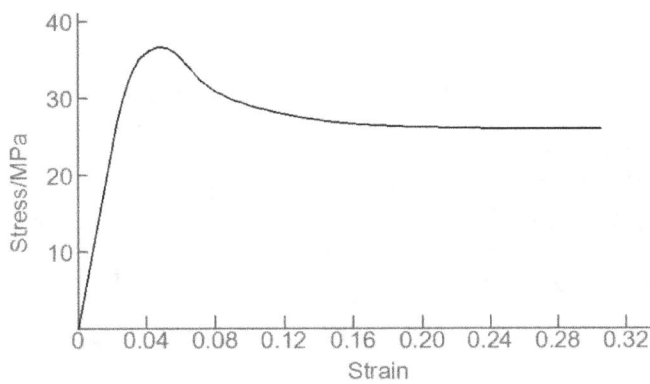

Figure 3.9

Answer:
  (a) The tensile modulus is the slope of the proportional part of the stress-strain graph and is thus about 28/0.02 = 1400 MPa = 1 . 4 GPa.
  (b) The tensile strength is the maximum stress. This is about 34 MPa.

Example
Referring to Figure 3.9, of the stress-strain graph for a sample of Novodur by Bayer (UK) Ltd ABS resin; estimate:
  (a) the tensile modulus
  (b) the tensile strength.

Answer:
  (a) The tensile modulus is the slope of the proportional part of the stress-strain graph and is thus about 28/0.03 = 933 MPa = 0.9 GPa.
  (b) The tensile strength is the maximum stress. This is about 34 MPa.

Bend Tests:

A simple test that is often quoted by suppliers of materials as a measure of ductility is the bend test which involves bending a sample of the material through some angle and determining whether the material is unbroken and free from cracks after the bending. There are a number of ways that can be used to carry out such a test using AS1580 and AS2505. The simplest method is the mandrel (a metal rod or bar) form of test shown in Figure 3.10, this being suitable for medium and thin thickness sheet for angles of bend up to 120°. Figure 3.11 shows how the test can be conducted on a vee block, this being suitable for medium thickness sheet with bend angles up to 90°. Figure 3.12 illustrates the form of test possible for thin sheet with bend angles up to 90°, the material being bent on a block of soft material. Other methods can also be used, e.g. bending round a mandrel, free bending and pressure bending (see the Australian Standards for more details). The results of a bend test are quoted in terms of the angle of bend that can be withstood without breaking or cracking, as illustrated in Figure 3.13.

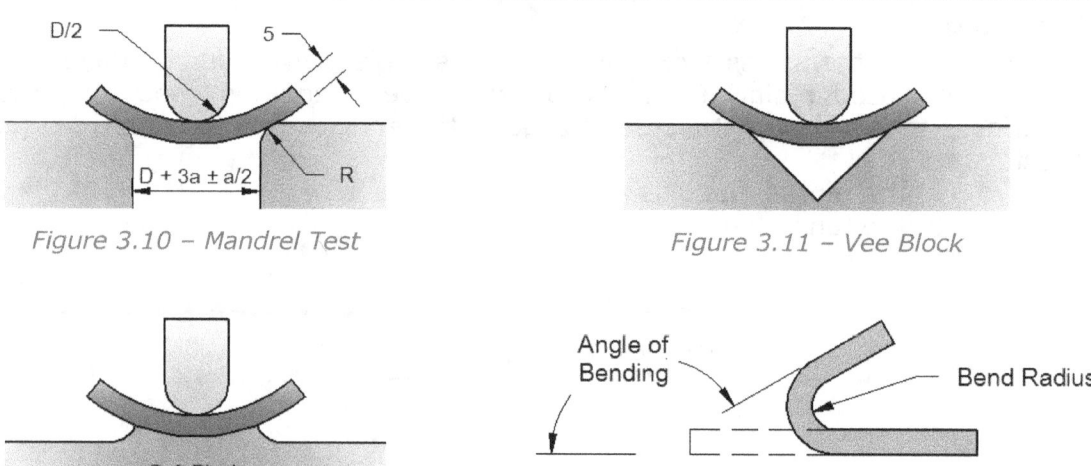

Figure 3.10 – Mandrel Test

Figure 3.11 – Vee Block

Figure 3.12 – Soft Block

Figure 3.13 – Angle of Bend

Impact Tests:

Impact tests are designed to simulate the response of a material to a high rate of loading and involve a test piece being struck a sudden blow. There are two main forms of test: the *Izod* and *Charpy* tests; both involve the same type of measurement but differ in the form of the test pieces; both involve a pendulum swinging down from a specified height h_o to strike the test piece as shown in Figure 3.14. The height h to which the pendulum rises after striking and breaking the test piece is a measure of the energy used in the breaking. If no energy were used the pendulum would swing up to the same height h_o as it started from, i.e. the potential energy mgh_o at the top of the pendulum swing before and after the collision would be the same. The greater the energy used in the breaking, the lower the height to which the pendulum rises. If the pendulum swings up to a height h after breaking the test piece then the energy used to break it is $mgh_o - mgh$.

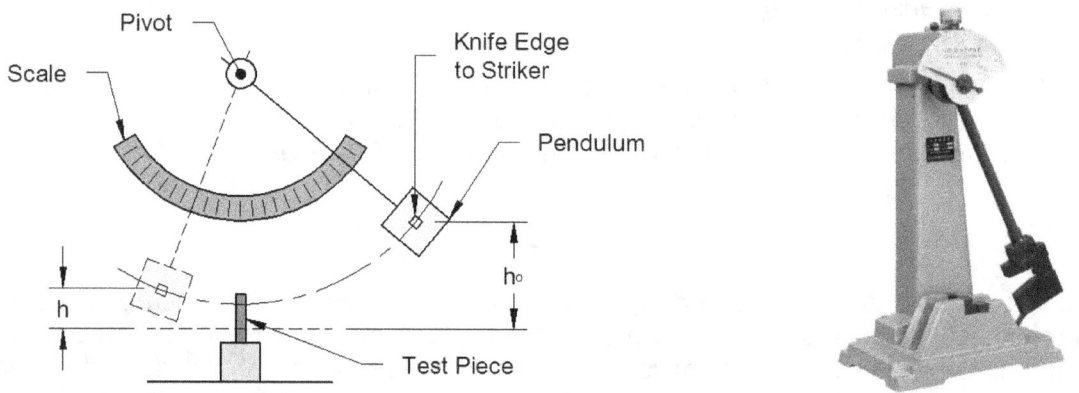

Figure 3.14 – Principle of the Impact Test

Figure 3.15 – Impact Test Equipment

Figure 3.14 is an example of a typical impact testing machine.

Izod V-Notch Test Pieces:

With the Izod test, the energy absorbed in breaking a cantilevered test piece is measured. The test piece has a notch and the blow is struck on the same face as the notch and at a fixed height above it.

Izod impact tests are setup and carried out in accordance with AS1544.1-2003. Test pieces are made from 10 mm x10 mm, 7.5 mm x 7.5 mm square section, 10mm x 5mm rectangular section, or 11.43mm diameter round section. The number of notches is dependent on the length of the test piece, 70mm long lengths require 1 notch (Figure 3.16), 98 mm lengths have 2 notches on opposite faces (Figure 3.17), while 126 mm lengths have 3 notches on separate faces (Figure 3.18).

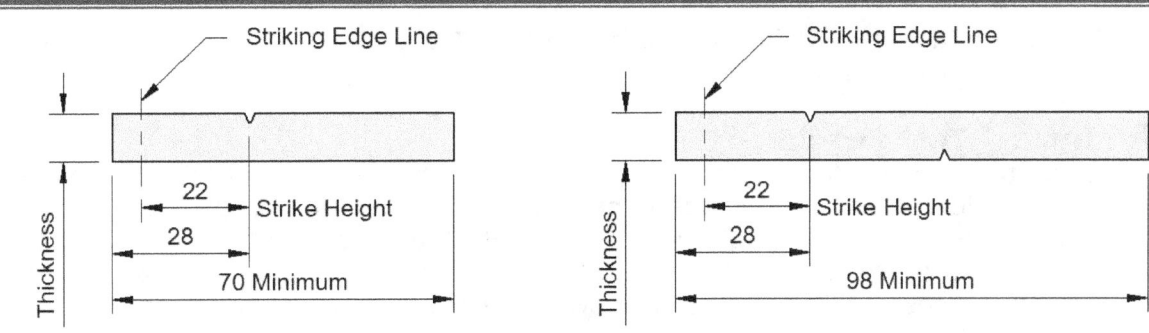

Figure 3.16 – Single Notch Metal Test Piece   Figure 3.17 – Double Notch Metal Test Piece

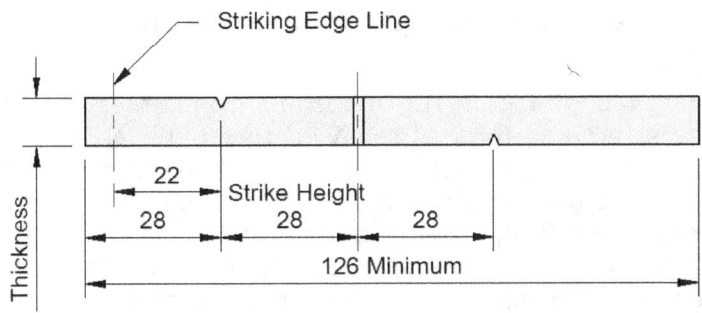   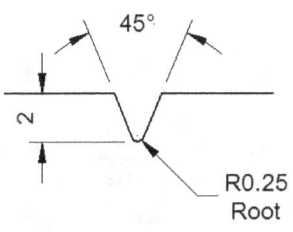

Figure 3.18 – Triple Notch Metal Test Piece   Figure 3.19 – Notch Detail

The notch detail must be in accordance with Figure 3.21. The pendulum strikes metal test pieces with a speed of between 3 and 4 m/s.

Charpy V-Notch Test Pieces:
The Charpy V-notch Impact Test is a pendulum-type single-blow impact test in which the test piece, V-notched in the middle and supported at both ends as a simple beam, is broken by a falling pendulum which strikes the test piece opposite the notch. The energy absorbed is determined from the subsequent rise of the pendulum The Australian Standard for the setup and administration of the Charpy impact test is in accordance with AS1544.2-2003.

Each standard test piece is 55 mm x 10 mm x 10 mm square section. Subsidiary test pieces are the 55mm long but either 10 mm x 7.5mm, 10 mm x 5mm or 10mm x 2.5mm rectangular sections.

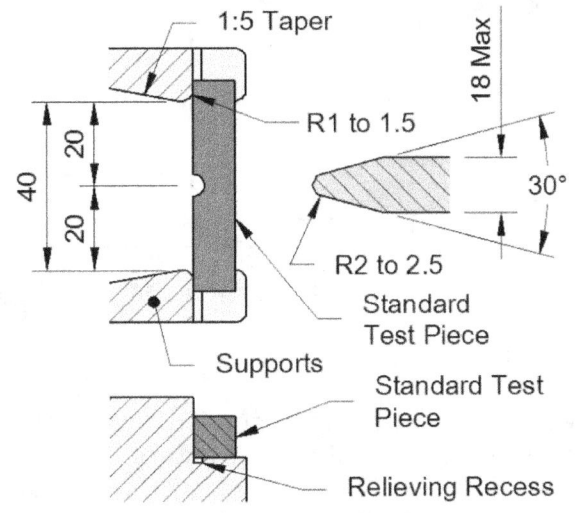   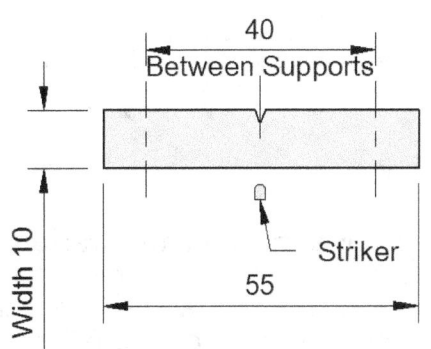

Figure 3.20 – Charpy Arrangement of Supports, Test Piece and Striker   Figure 3.21 – Charpy Test Piece

The notch detail must be in accordance with Figure 3.21. The striker velocity is 4.5 to 7.0 m/s.

Impact Test Results:

In stating the results of impact tests it is vital that the form of test is specified; there is no reliable relationship between the values obtained by the two forms of test and so values from one test cannot be compared with those from the other. In addition, there is no reliable relationship between the impact energies given for breaking test pieces of different sizes or different notches with the same test method. The impact energy is influenced by such factors as the temperature, the speed of impact, any degree of directionality in the properties of the material from which the test piece was cut, and the thickness of the test piece. For both the Izod and Charpy tests, the impact strengths for metals are expressed in the form of the energy absorbed; for example, 30 J.

Interpreting Impact Test Results:

When a material is stretched, energy is stored in the material. Similar to stretching a spring or a rubber band, when the stretching force is released the material springs back and the energy is released. However, if the material suffers a permanent deformation then all the energy is not released. The greater the amount of such plastic deformation, the greater the energy not released, therefore, when a ductile material is broken, more energy is 'lost'. The fracture of materials can be classified roughly as either brittle or ductile. With brittle fracture there is little plastic deformation prior to fracture and so little energy is required to break the test piece. With ductile fracture the fracture is preceded by a considerable amount of plastic deformation and so more energy is required to break the test piece, thus the impact test can be used to give information about the type of fracture that occurs. For example Figure 3.22 shows the effect of temperature on the Charpy V-notch impact energies obtained for test pieces of 0.2% carbon steel; above about 0°C the material gives ductile failures, below that temperature, brittle failures. Such graphs have a great bearing on the use that can be made of the material, since at low temperatures the steel can be easily shattered by impact.

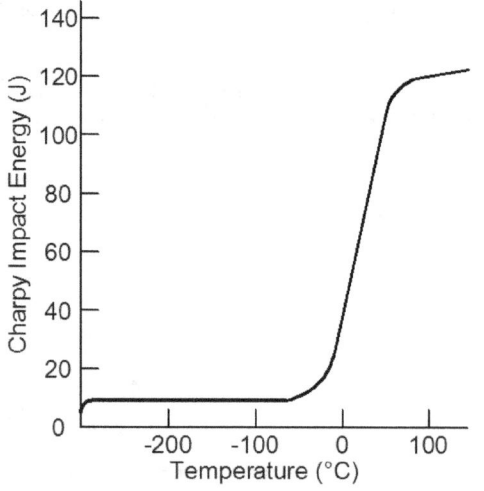

Figure 3.22 – Single Notch Metal Test Piece   Figure 3.23 – Double Notch Metal Test Piece

The appearance of the fractured surfaces after an impact test also gives information about the type of fracture that has occurred. With a brittle fracture of metals, the surfaces are crystalline in appearance while in a ductile fracture, they are rough and fibrous; also with ductile failure there is a significant reduction in the cross-sectional area of the test piece, but with brittle fracture there is virtually no such change.

One use of impact tests is to determine whether heat treatment has been successfully carried out. A comparatively small change in heat treatment can lead to quite noticeable changes in impact test results; these can be more pronounced than changes in other mechanical properties such as percentage elongation or tensile strength. Figure 3.23 shows the effect of annealing to different temperatures on the Izod impact test results for cold worked mild steel. The impact test can therefore be used to indicate whether annealing has been carried out to the required temperature.

Toughness Test:
Fracture toughness testing involves test pieces with sharp notches being strained until the crack propagates and the test piece fails. The problem in obtaining test pieces is producing the sharp notches; this is done by taking a test piece with a machined notch and then using a standardized pre-cracking procedure by loading with an alternating stress (fatigue loading) in order to obtain a sharp crack at the base of the machined notch.

Hardness Tests:
Hardness is the property of a material that enables it to resist plastic deformation, usually by penetration; however, the term hardness may also refer to resistance to bending, scratching, abrasion or cutting. Hardness is not an intrinsic material property dictated by precise definitions in terms of fundamental units of mass, length and time. A hardness property value is the result of a defined measurement procedure.

Hardness of materials has probably long been assessed by resistance to scratching or cutting. An example would be material B scratches material C, but not material A. Alternatively, material A scratches material B slightly and scratches material C heavily. Relative hardness of minerals can be assessed by reference to the Moh Scale that ranks the ability of materials to resist scratching by another material. Similar methods of relative hardness assessment are still commonly used today; a simple example is the file test where a file tempered to a desired hardness is rubbed on the test material surface. If the file slides without biting or marking the surface, the test material would be considered harder than the file. If the file bites or marks the surface, the test material would be considered softer than the file.

The above relative hardness tests are limited in practical use and do not provide accurate numeric data or scales particularly for modern day metals and materials. The usual method to achieve a hardness value is to measure the depth or area of an indentation left by an indenter of a specific shape, with a specific force applied for a specific time. Three principal standard test methods can be used for expressing the relationship between hardness and the size of the impression, these being Brinell, Vickers, and Rockwell; for practical and calibration reasons, each of these methods is divided into a range of scales, defined by a combination of applied load and indenter geometry.

The Brinell Hardness Test:
The Brinell hardness test method consists of indenting the test material with a 10 mm diameter hardened steel or carbide ball subjected to a load of 3000 kg. For softer materials the load can be reduced to 1500 kg or 500 kg to avoid excessive indentation. The full load is normally applied for 10 to 15 seconds in the case of iron and steel and for at least 30 seconds in the case of other metals. The diameter of the indentation left in the test material is measured with a low powered microscope. The Brinell harness number is calculated by dividing the load applied by the surface area of the indentation.

The diameter of the impression is the average of two readings at right angles and the use of a Brinell hardness number table can simplify the determination of the Brinell hardness. A well-structured Brinell hardness number reveals the test conditions, for example, *75 HB 10/500/30* which means that a Brinell Hardness of 75 was obtained using a 10mm diameter hardened steel with a 500 kilogram load applied for a period of 30 seconds. On tests of extremely hard metals a tungsten carbide ball is substituted for the steel ball. Compared to the other hardness test methods, the Brinell ball makes the deepest and widest indentation, so the test averages the hardness over a wider amount of material, which will more accurately account for multiple grain structures and any irregularities in the uniformity of the material; this method is the best for achieving the bulk or macro-hardness of a material, particularly those materials with heterogeneous structures.

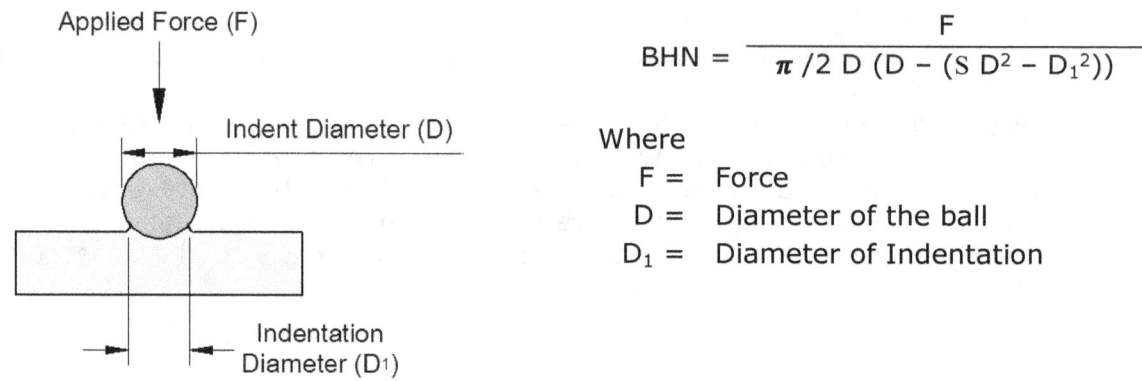

$$BHN = \frac{F}{\pi/2\ D\ (D - (S D^2 - D_1^2))}$$

Where
F = Force
D = Diameter of the ball
D_1 = Diameter of Indentation

The units used for the area are mm² and for the force kgf (1 kgf = 9.8 N and is the gravitational force exerted by 1 kg). The area can be obtained, from the measured diameter of the indentation and ball diameter, by either calculation or the use of tables.

The Brinell test cannot be used with very soft or very hard materials. In the one case the indentation becomes equal to the diameter of the ball and in the other there is either no or little indentation on which measurements can be based. The thickness of the material being tested should be at least ten times the depth of the indentation if the results are not to be affected by the thickness of the material.

The Vickers Hardness Test:

The Vickers hardness test method consists of indenting the test material with a diamond indenter, in the form of a right pyramid with a square base and an angle of 136 degrees between opposite faces subjected to a load of 1 to 100 kgf. The full load is normally applied for 10 to 15 seconds. The two diagonals of the indentation left in the surface of the material after removal of the load are measured using a microscope and their average calculated. The area of the sloping surface of the indentation is calculated. The Vickers hardness is the quotient obtained by dividing the kgf load by the square mm area of indentation.

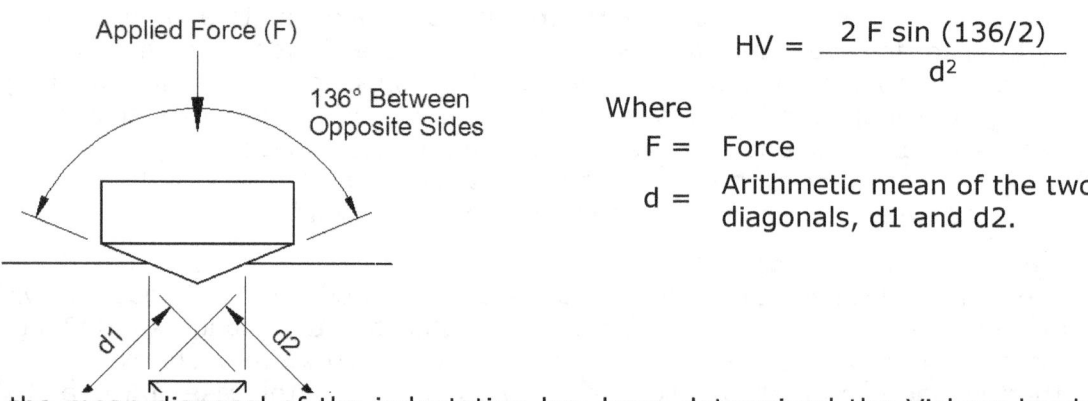

$$HV = \frac{2 F \sin(136/2)}{d^2}$$

Where
F = Force
d = Arithmetic mean of the two diagonals, d1 and d2.

When the mean diagonal of the indentation has been determined the Vickers hardness may be calculated from the formula, but is more convenient to use conversion tables. The Vickers hardness should be reported like *800 HV/10*, which means a Vickers hardness of 800, was obtained using a 10 kgf force. Several different loading settings give practically identical hardness numbers on uniform material, which is much better than the arbitrary changing of scale with the other hardness testing methods. The advantages of the Vickers hardness test are that extremely accurate readings can be taken, and just one type of indenter is used for all types of metals and surface treatments. Although thoroughly adaptable and very precise for testing the softest and hardest of materials, under varying loads, the Vickers machine is a floor standing unit that is more expensive than the Brinell or Rockwell machines.

The Rockwell Hardness Test:

The Rockwell hardness test is governed by Australian Standard AS1815.1. The test consists of indenting the test material with a diamond cone or hardened steel ball indenter. The indenter is forced into the test material under a preliminary minor load *F0* (Load A) usually 10 kgf. When equilibrium has been reached, an indicating device, which follows the movements of the indenter and so responds to changes in depth of penetration of the indenter, is set to a datum position. While the preliminary minor load is still applied an additional major load is applied with resulting increase in penetration (Load B); when equilibrium has again been reach, the additional major load is removed but the preliminary minor load is still maintained. Removal of the additional major load allows a partial recovery, so reducing the depth of penetration (Load C). The permanent increase in depth of penetration, resulting from the application and removal of the additional major load is used to calculate the Rockwell hardness number.

$$HR = E - e$$

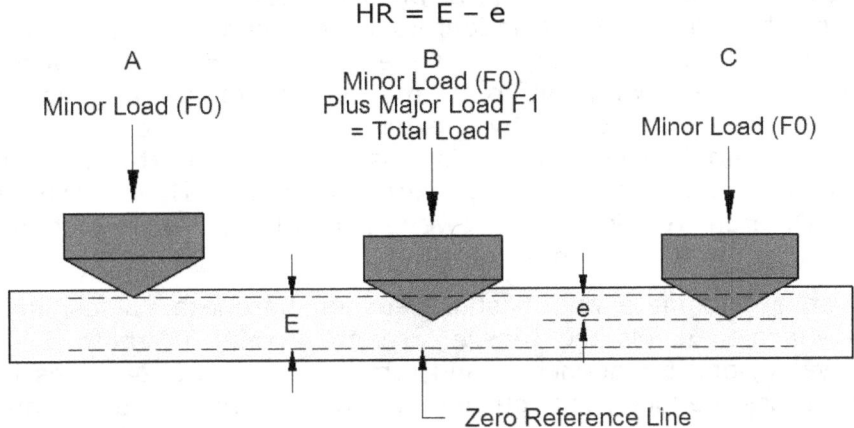

- F = Total load in kgf.
- $F0$ = Preliminary load in kgf.
- $F1$ = Additional major load in kgf.
- e = Permanent increase in the depth penetration due to major load $F1$ measured in units of 0.02 mm.
- E = A constant depending on form of indenter: 100 units for diamond indenter, 130 units for steel ball indenter.
- HR = Rockwell hardness number
- D = diameter of steel ball

Rockwell Hardness Scale	Hardness Symbol	Type of Indenter	Preliminary Test Force F0 N	Additional Test Force F1 N	Total Test Force F N	Field of Application
A	HRA	Diamond Cone	98.07	490.3	588.4	20 HRA to 88 HRA
B	HRB	Ball 1.5875 mm	98.07	882.6	980.7	20 HRB to 100 HRB
C	HRC	Diamond Cone	98.07	1373	1471	20 HRC to 70 HRC
D	HRD	Diamond Cone	98.07	882.6	980.7	40 HRD to 77 HRD
E	HRE	Ball 3.175 mm	98.07	882.6	980.7	70 HRE to 100 HRE

F	HRF	Ball 1.5875 mm	98.07	490.3	588.4	60 HRF to 100 HRF
G	HRG	Ball 1.5875 mm	98.07	1373	1471	30 HRG to 94 HRG
H	HRH	Ball 3.175 mm	98.07	490.3	588.4	80 HRH to 100 HRH
K	HRK	Ball 3.175 mm	98.07	1373	1471	40 HRK to 100 HRK
15N	HR15N	Diamond Cone	29.42	117.7	1471	70 HRG to 94 HR15N
30N	HR30N	Diamond Cone	29.42	264.8	294.2	42 HR30N to 86 HR30N
45N	HR45N	Diamond Cone	29.42	411.9	441.3	20 HR45N to 77 HR45N
15T	HR15T	Ball 1.5875 mm	29.42	117.7	147.1	67 HR15T to 93 HR15T
30T	HR30T	Ball 1.5875 mm	29.42	264.8	294.2	29 HR30T to 82 HR30T
45T	HR45T	Ball 1.5875 mm	29.42	411.9	441.3	10 HR45T to 72 HR45T

Comparison of the Different Hardness Scales:

The Brinell and Vickers tests both involve measurements of the surface area of indentations, the forms of the indenters used being different while the Rockwell test entails measurements of the depth of penetration of indenters. Consequently the various tests are concerned with different measurements as an indication of hardness. Consequently the values given by the different methods differ for the same material. There are no simple theoretical relationships between the various hardness scales, though some simple approximate, experimentally derived, relationships have been obtained. Different relationships, however, hold for different metals. The relationships are often presented in the form of tables.

There is an approximate relationship between hardness values and tensile strengths. Thus for annealed steels the tensile strength in MPa is about 3.54 times the Brinell hardness value, and for quenched and tempered steels 3.24 times the Brinell hardness value. For brass the factor is about 5.6 and for aluminium alloys about 4.2.

The Moh Scale of Hardness:

The Moh hardness scale for minerals has been used since 1822. It simply consists of 10 minerals arranged in order from 1 to 10. Diamond is rated as the hardest and is indexed as 10; talc as the softest with index number 1. Each mineral in the scale will scratch all those below it as follows:

Diamond	10	Apatite	5
Corundum	9	Fluride	4
Topaz	8	Calcite	3
Quartz	7	Gypsum	2
Orthoclase (Feldspar)	6	Talc	1

The steps are not of equal value and the difference in hardness between 9 and 10 is much greater than between 1 and 2. The hardness is determined by finding which of the standard minerals the test material will scratch or not scratch; the hardness will lie between two points on the scale - the first point being the mineral which is scratched and the next point being the mineral which is not scratched. Some examples of the hardness of common metals in the Moh scale are copper between 2 and 3 and tool steel between 7 and 8. This is a simple test, but is not exactly quantitative and the standards are purely arbitrary numbers.

The materials engineer and metallurgist find little use for the Moh scale, but it is possible to sub-divide the scale and some derived methods are still commonly used today. The file test is useful as a rapid and portable qualitative test for hardened steels, where convention hardness testers are not available or practical. Files can be tempered back to give a range of known hardness and then used in a similar fashion to the Moh method to evaluate hardness.

Hardness Values:

Figure 3.24 shows the general range of hardness values for different types of material when measured by the Vickers, Brinell, Rockwell and Moh test methods.

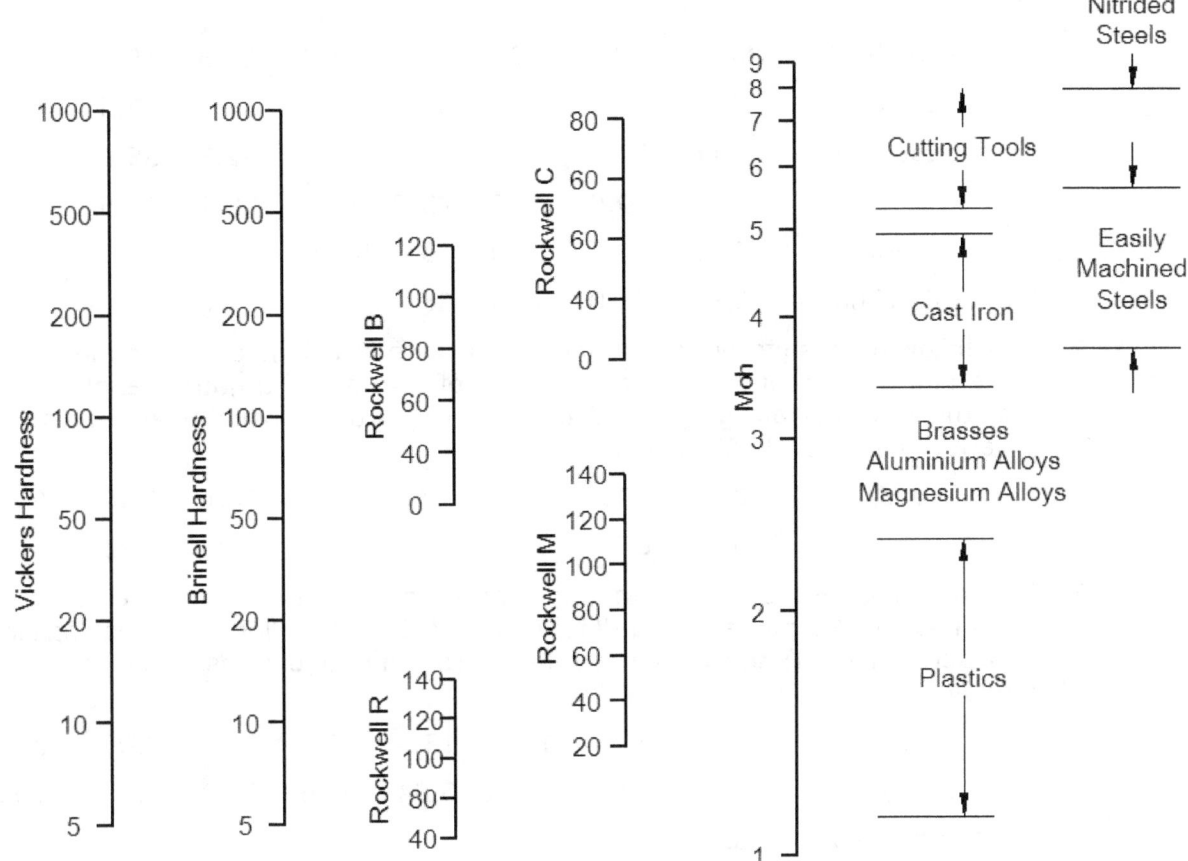

Figure 3.24 – Hardness Values

MEM30007A - Select common engineering materials
Topic 3 - Materials Testing

Review Problems:
MEM30007-RQ-03

1. The following results were obtained from a tensile test of an aluminium alloy. The test piece had a diameter of 11.28 mm and a gauge length of 56 mm. Plot the stress-strain graph and determine (a) the tensile modulus and (b) the 0.1% proof stress.

Load/kN	0	2.5	5.0	7.5	10.0	12.5	15.0	17.5
Ext./mm	0	1.8	4.0	6.2	8.4	10.0	12.5	14.6
Load/kN	20.0	22.5	25.0	27.5	30.0	32.5	35.0	
Ext./mm	16.3	19.0	21.2	23.5	25.7	28.1	31.5	
Load/kN	37.5	38.5	39.0	39.0 (broke)				
Ext./mm	35.0	40.0	61.0	86				

2. The following results were obtained from a tensile test of a polymer. The test piece had a width of 20 mm, a thickness of 3 mm and a gauge length of 80 mm. Plot the stress-strain graph and determine (a) the tensile strength and (b) the secant modulus at 0.2% strain.

Load/N	0	100	200	300	400	500	600	650	630
Ext./mm	0	0.08	0.17	0.35	0.59	0.88	1.33	2.00	2.40

3. The following results were obtained from a tensile test of steel. The test piece had a diameter of 10 mm and a gauge length of 50 mm. Plot the stress-strain graph and determine (a) the tensile strength; (b) the yield stress and (c) the tensile modulus.

Load/kN	0	5	10	15	20	25	30	32.5
Ext./mm	0	0.016	0.033	0.049	0.065	0.081	0.097	0.106
Load/kN	35.8							
Ext./mm	0.250							

4. A flat tensile test piece of steel has a gauge length of 100.0 mm. After fracture, the gauge length was 131.1 mm. What is the percentage elongation?

5. The following data were obtained from a tensile test on a stainless steel test piece. Determine (a) the limit of proportionality stress; (b) the tensile modulus and (c) the 0.2% proof stress.

Stress/MPa	0	90	170	255	345	495	605
Strain/x10^{-4}	0	5	10	15	20	30	40
Stress/MPa	700	760	805	845	880	895	
Strain/x10^{-4}	50	60	70	80	90	100	

6. Estimate from the stress-strain graph for cast iron in Figure 3.25 the tensile strength and the limit of proportionality.

7. Estimate from the stress-strain graph for a sample of nylon 6 given in Figure 3.26 the tensile modulus and the tensile strength.

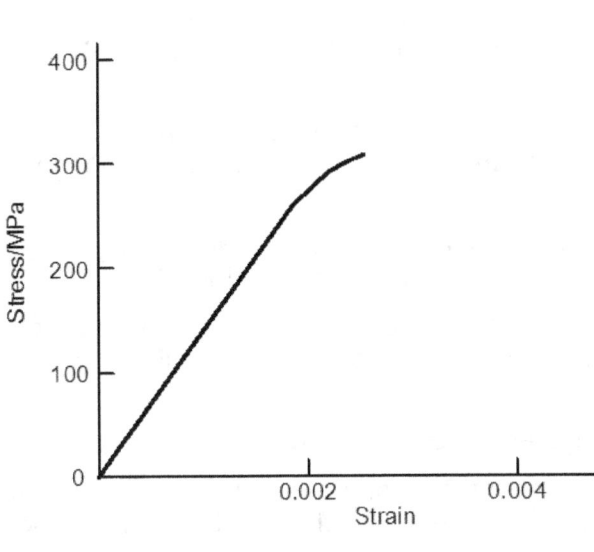
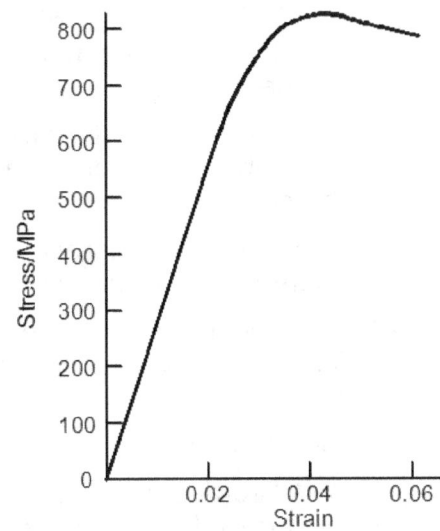

Figure 3.25   Figure 3.26

8. Sketch the form of the stress-strain graphs for (a) brittle stiff materials; (b) brittle non-stiff materials; (c) ductile stiff materials and (d) ductile non-stiff materials.

9. The effect of working an aluminium alloy (1.25% Mn) is to change the tensile strength from 110 MPa to 180 MPa and the elongation from 3 0% to 3%. What is the effect of the working on the properties of the material?

10. An annealed titanium alloy has a tensile strength of 880 MPa and an elongation of 16%. An annealed nickel alloy has a tensile strength of 700 MPa and an elongation of 35%. Which alloy is (a) the stronger and (b) the more ductile in the annealed condition?

11. Cellulose acetate has a tensile modulus of 1.5 GPa and polythene a modulus of 0.6 GPa. Which of the two plastics will be the stiffer?

12. The following are Izod impact energies at different temperatures for samples of annealed cartridge brass (7 0% C u - 3 0% Zn). What can be deduced from the results?

Temperature (°C)	+27	-78	-197
Impact energy (J)	88	92	108

13. The following are Charpy V-notch impact energies for annealed titanium at different temperatures. What can be deduced from the results?

Temperature (°C)	+27	-78	-196
Impact energy (J)	24	19	15

14. The following are Charpy impact strengths for nylon 6.6 at different temperatures. What can be deduced from the results?

Temperature (°C)	-23	-33	-43	-63
Impact strength (kJ/m^2)	24	13	11	8

15. The impact strengths of samples of nylon 6, at a temperature of 22°C, are found to be 3 kJ/m^2 in the as-moulded condition but 25 kJ/m^2 when the sample has gained 2.5% in weight through water absorption. What can be deduced from the results?

16. With the Vickers hardness test a 30 kg load gave for a sample of steel an indentation with diagonals having mean lengths of 0.530 mm. What is the hardness?

17. With the Vickers hardness test a 30 kg load gave for a sample of steel an indention with diagonals having mean lengths of 0.450 mm. What is the hardness?

18. With the Vickers hardness test a 10 kg load gave for a sample of brass an indentation with diagonals having mean lengths of 0.510 mm. What is the hardness?

19. With the Brinell hardness test a 10mm diameter ball and 3000 kg load gave an indentation with a diameter of 4.10 mm. What is the hardness?

20. With the Brinell hardness test a sample of cold-worked copper with a 1 mm diameter ball and 20 kg load gave an indentation of diameter 0.630 mm. What is the hardness?

21. Specify the type of test that can be used in the following instances:

    a) A storekeeper has mixed up two batches of steel, one batch having been surface hardened and the other not. How could the two be distinguished?

    b) What test could be used to check whether tempering has been correctly carried out for a steel?

    c) A plastic is modified by the inclusion of glass fibres. What test can be used to determine whether this has made the plastic stiffer?

    d) What test could be used to determine whether a metal has been correctly heat treated?

    e) What test could be used to determine whether a metal is in a suitable condition for forming by bending?

Topic 4 – Structure and Properties:

Required Skills:
- Discuss the basic structure of metals, polymers and composites and the factors which affect the structure.
- Explain changes in properties in terms of changes in structure.
- Associate mechanical properties with particular structures.
- Identify structures which can lead to required properties.

Required Knowledge:
- Various types of metals.
- Reading graphs, tables and charts.
- Testing procedures.
- Processing formulae.

Structure of Metals:

Metals are giant structures of atoms held together by metallic bonds. "*Giant*" implies that large but variable numbers of atoms are involved – depending on the size of the bit of metal.

The term "*metal*" is used for elements such as copper, which are made up of atoms. In the solid state, copper consists of an array of atoms each of which has lost one electron (Figure 4.1) leaving each copper atom as having a net positive charge, termed a positive ion. The electrons that have been lost do not combine with any one ion but remain as a cloud of negative charge floating between the ions. The result is rather like a glue in that the cloud of electrons holds the positive ions together.

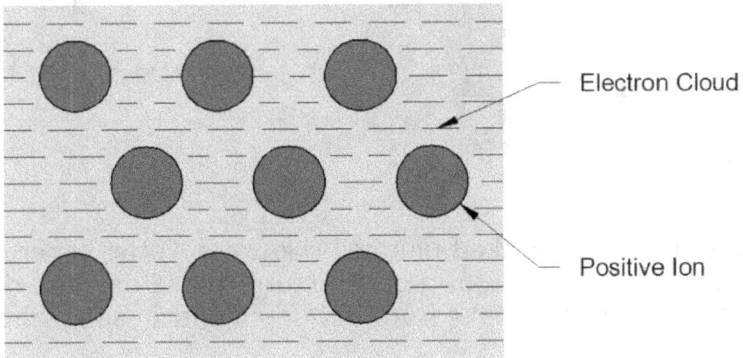

Figure 4.1

A simple model used to describe the structure of metals is to think of the ions in a metal as spheres. Since the bonds formed between the positive ions can be formed in any direction without any restrictions, the only rule on how the spheres can be arranged is that imposed by how the geometry of the sphere restricts the packing together of spheres in order to give a tightly packed structure.

The free electrons explain why metals are good conductors of electricity, since they have free charge carriers which are easily moved through the solid by the application of a voltage. Insulators have no free electrons and the atoms in the solid are bonded together in a different way.

Crystals:

It would be misleading to suppose that all the atoms in a piece of metal are arranged in a regular way. Any piece of metal is made up of a large number of "*crystal grains*", which are regions of regularity. At the grain boundaries atoms have become misaligned.

One of the simplest arrangement of spheres is that of the simple cubic structure. Figure 4.2 shows the structure obtained by stacking four spheres with the centres of each sphere at the corners of a cube. The surfaces of each sphere touch the surfaces of each of its neighbours in such a way that the length of the side of the cube is equal to the diameter of the spheres.

The dotted line in the Figure encloses what is termed the unit cell. To form a solid we can consider that this structure is repeated many times. The resulting solid would consist of a completely orderly array of spheres, i.e. atoms. We would expect the surfaces of such a solid to be smooth and flat with the angles between adjoining faces always 90°. Such a solid would, when broken up, always have the appearance of stacked cubes. This is a description of a cubic crystal.

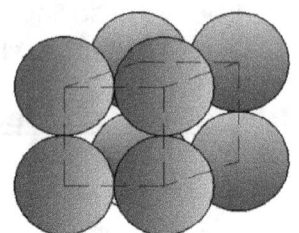

Figure 4.2 – Simple Cubic Structure

A crystal consequently consists of a large number of particles arranged in a regular repetitive array; it is this regularity which is characteristic of crystalline material. A solid having no such order in the arrangement of its constituent particles is said to be amorphous.

The simple cubic crystal shape is arrived at by stacking spheres in one particular way; it is not the way that spheres can be most closely packed. By stacking spheres in a closer manner, as shown in the following Figures, other crystal shapes can be produced. With the body-centred cubic unit cell (Figure 4.3) the arrangement is slightly more complex than the simple cubic unit cell in having an extra sphere in the centre of the cell.

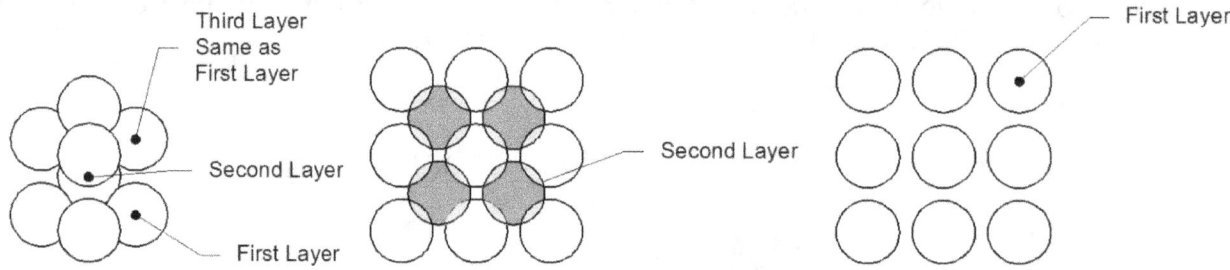

Figure 4.3 – Body Centred Cubic Structure

With the Hexagonal Close-Packed unit cell (Figure 4.4) the spheres are packed in a close array which gives a hexagonal form of structure; these are just the three closest packed arrangements which we can make from packing identical spheres together

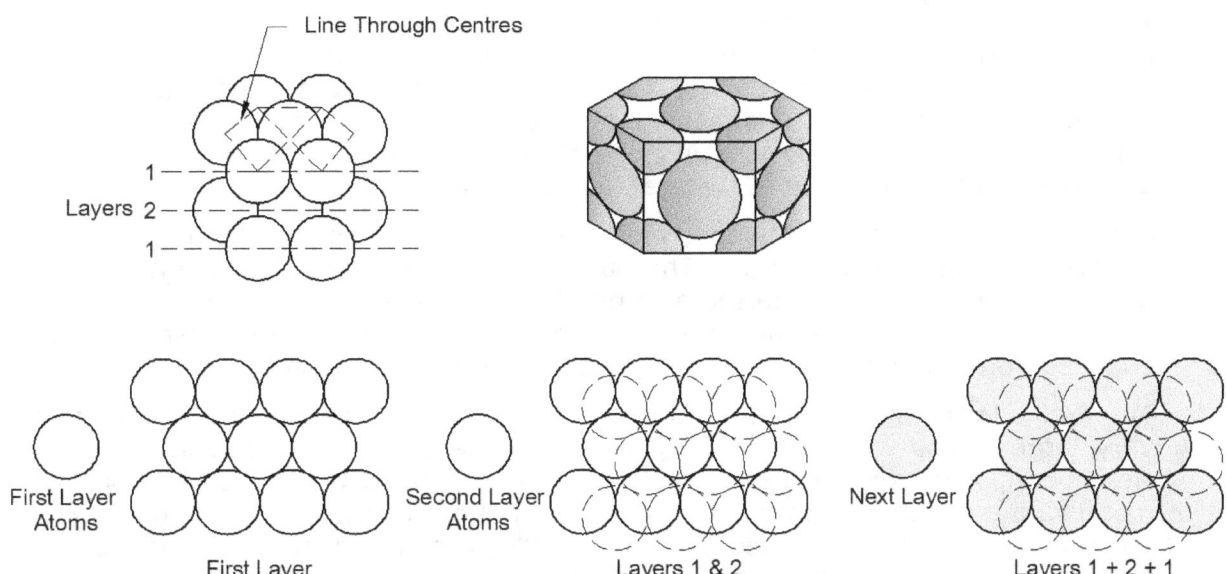

Figure 4.4 – Hexagonal Close-Packed Structure

With the **Face-Centred Cubic** unit cell (Figure 4.5) there is, when compared with the simple cubic unit cell, a sphere at the centre of each face of the cube.

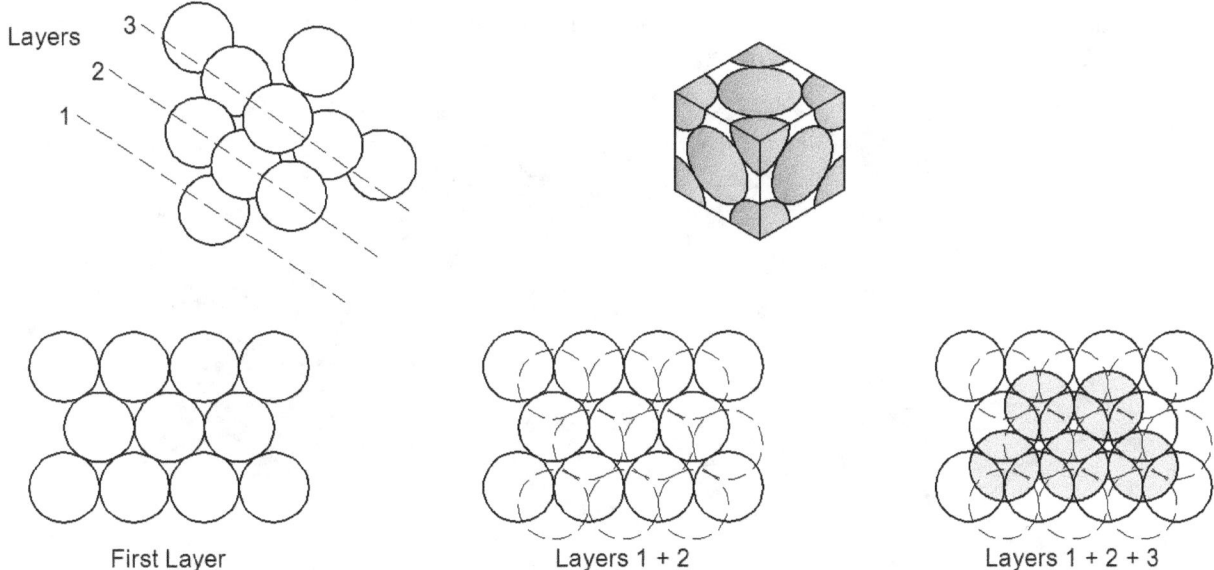

Figure 4.5 – Face-Centred Cube Structure

Because metals can be considered to be composed of spherical atoms, it is these three close-packed structures which are used for solid metals.

An important point to note with all these structures is that there are spaces between the spheres in the crystal structures. The size of these spaces depends on the type of structure. Within these spaces it is possible to fit other atoms, provided they are small enough, without too much strain on the crystalline structure. In some cases, with some strain, atoms can be forced into spaces which are really too small for them; this is discussed later in the chapter in connection with alloys.

Crystalline Structure:
Metals are crystalline substances. The atoms in crystals are packed together in a regular, repeating pattern, known as the space lattice: solids with no such order are said to be "*amorphous*". The unit cell is the simplest geometric figure which represents the grouping of particles: a crystal consists of a large number of these unit cells stacked together as shown in Figure 4.3, Figure 4.4 and Figure 4.5.

There are many other crystal structures found in metals, but attention is concentrated on particular close-packed structures because they are the crystal forms found in the metals used for engineering purposes. This is because close-packed structures are linked with ductility, where materials under load start stretching before they break; the movement can be used to give early warning of potential failure, and is to be preferred to the alternative of "*brittle fracture*", where there is no similar indication.

A simple model of a metal with grains is given if a raft of bubbles is produced on the surface of a liquid (Figure 4.5). The bubbles pack together in an orderly and repetitive manner but if "*growth*" is started at a number of centres then "*grains*" are produced. At the boundaries between the grains the regular pattern breaks down as the pattern changes from the orderly one of one grain to that of the next grain.

The grains in the surface of a metal are not generally visible, though an exception is the very large grains which are readily visible in the surface of galvanized steel objects. Grains can, however, be made visible by careful etching of the polished surface of the metal with a suitable chemical. The chemical preferentially attacks the grain boundaries. For example, in the case of copper and its alloys, concentrated nitric acid can be used. In the case of carbon and alloy steels of medium carbon content an etchant called nital can be used. Nital is a mixture of nitric acid and alcohol, typically 5 ml of acid to 95 ml of alcohol.

Examples of metals which, in the pure state, adopt the body-centred cubic unit cell form of structure are iron, chromium and molybdenum, face-centred cubic unit cell forms of structure are aluminium, copper, lead and nickel, with the hexagonal close-packed unit cell being given by magnesium and zinc.

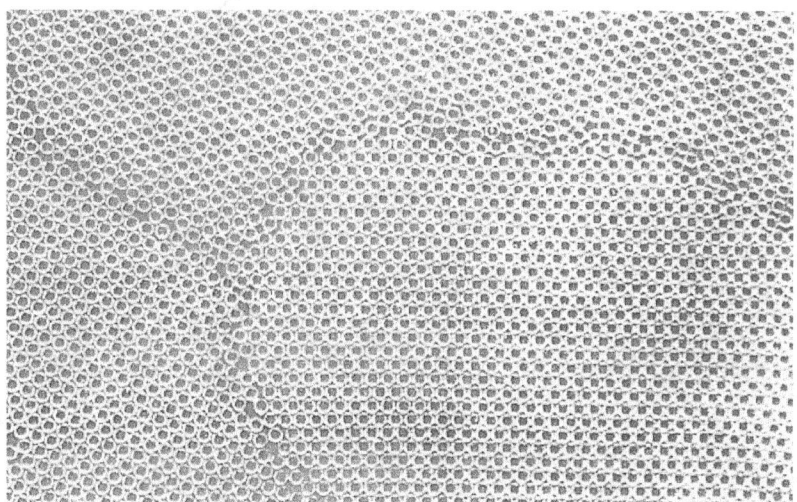

Figure 4.6 – Grains in a Bubble Raft

Alloys:

An alloy is a metal made by combining two or more metallic element, to give greater strength or resistance to corrosion. Most metallic components used in industry, commerce, to motor vehicles and machinery are manufactured from alloys rather than the pure metals. Pure metals do not always have the appropriate combinations of properties needed while alloys can be designed and manufactured to specifically required specifications.

Making alloys is similar to baking a cake. The basic ingredients of flour, sugar, butter, eggs and water are mixed together and then cooked. The result is a cake which has a texture and properties quite different from those of the individual ingredients. The type of cake produced depends on the relative amounts of the ingredients and the way it is cooked. In making alloys, the ingredients are mixed and heated and the resulting alloy can have properties quite different from those of the ingredients. The properties will depend on the relative amounts and nature of the ingredients as well as how they are 'baked'. An alloy is a particular mixture of components and so has a particular chemical composition, e.g. carbon steel may be 99.0% iron combined with 1.0% carbon while a corrosion resistant steel is 72% iron with 15% carbon and 13% chromium.

Australian coins are made of alloys. Coins need to be made of a relatively hard material which does not wear away rapidly, i.e. they must have a life of many years. If the coins were made of pure copper, they would be very soft; not only would they suffer considerable wear but they would bend trouser or purse pockets. The 5c, 10, 20c and 50c coins are made using 75% copper and 25% nickel while the $1 and $2 coins are made from 92% copper, 6% aluminium and 2% nickel. The formerly used 1c and 2c coins consisted of 97% copper, 2.5% zinc and 0.5% tin while the original round 50c coins was made from 80% silver and 20% copper.

Pure metals tend to be soft with high ductility, low tensile strength and low yield strength; because of this they are rarely used in engineering. Alloying can produce harder materials with higher tensile strength, higher yield stress and a reduction in ductility. Such materials are more useful in engineering; however, there are some circumstances in which the properties of pure metals are useful where high electrical conductivity is required (copper); where good corrosion resistance is required (lead); and where very high ductility is required.

Structure of alloys can be thought of in terms of the constituent metals, say A and B, being mixed in the liquid state; when the mixture solidifies, there is the possibility that solid alloy will have a crystal structure in which some of the atoms in the crystal structure of A have been replaced by atoms of B (Figure 4.7). Alternatively, because there are spaces between the atoms of A in its crystal structure, some atoms of A, if small enough, might lodge in these spaces (Figure 4.8). Another possibility is that elements A and B combine to form a chemical compound. With a compound there will be a particular structure for that compound with atoms of A and B assuming specific positions, rather than just popping into any gap. Another possibility is that when the liquid mixture cools A and B separate out, with B forming its own crystal structure independent of A. The structure then becomes a mixture of two types of crystals.

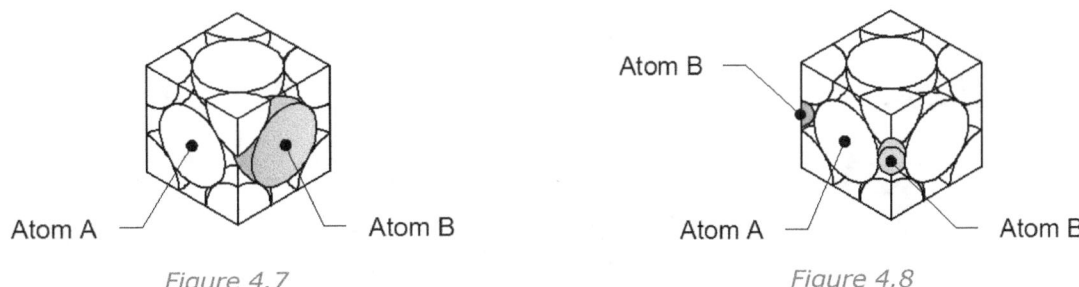

Figure 4.7   *Figure 4.8*

Ferrous Alloys:

Pure iron is a relatively soft material and seldom found in industry in its natural state; however, alloys of iron with carbon are very widely used. The term ferrous alloys is given to alloys with iron. Pure iron at room temperature exists as a body-centred cubic structure, and is commonly referred to as ferrite and continues to exist in this form for up to 912°C. At this temperature the structure changes to a face-centred cubic one, known as austenite. Iron atoms have a diameter of 0.256 nm (1 nanometre = 10^{-9} m), carbon atoms are much smaller with a diameter of 0.154 nm. The face-centred cubic structure is a more open structure than the body-centred cubic. The face-centred structure of austenite has voids which can accommodate spheres of up to 0.104 n min diameter; the body-centred cubic structure has voids between the atoms which are 0.070 nm in diameter, therefore, carbon atoms can be more easily accommodated within austenite, without severe distortion of the lattice, than ferrite. Austenite can take up to 2.0% of carbon while ferrite can take only 0.2%. When iron containing carbon is cooled from the austenite state to the ferrite state, there is a reduction in the amount of carbon that can be accommodated with in the iron and so some of the carbon atoms come out of the crystals and form a compound, another crystal structure, between iron and carbon called cementite which is hard and brittle. The result can be a structure consisting of purely ferrite grains mixed with grains which have a laminated structure of ferrite and cementite. Such a laminated structure is termed pearlite. Pure cementite is harder than pearlite, which in turn is harder than pure ferrite; consequently the structure, and the properties, of the iron alloy is determined by the amount of carbon present.

The percentage of carbon alloyed with iron has a profound effect on the properties of the alloy. The terms used for the alloys produced with different percentages of carbon are:
- Wrought iron   0 to 0.05% carbon
- Steel   0.05 to 2% carbon
- Cast iron   2 to 4.5% carbon

The term carbon steel is used for those steels in which essentially just iron and carbon are present. Alloy steel is used where other elements are included.

Plain Carbon Steel:

Carbon steels are grouped according to their carbon content with the designations being roughly:

Mild steel	0.10 to 0.25% carbon
Medium-carbon steel	0.20% to 0.50% carbon
High-carbon steel	More than 0.50% carbon

Mild steel has a structure consisting predominantly of ferrite, medium carbon steels tend to have about equal amounts of ferrite and pearlite, while high-carbon steels have predominantly pearlite with some free cementite occurring at high carbon contents. Figure 4.9 shows how the mechanical properties of carbon steels depend on the percentage of carbon. Increasing the percentage of carbon, within the range considered, increases the amount of pearlite at the expense of the softer ferrite, and hence:

1. Increases the tensile strength
2. Increases the hardness
3. Decreases the percentage elongation
4. Decreases the impact strength

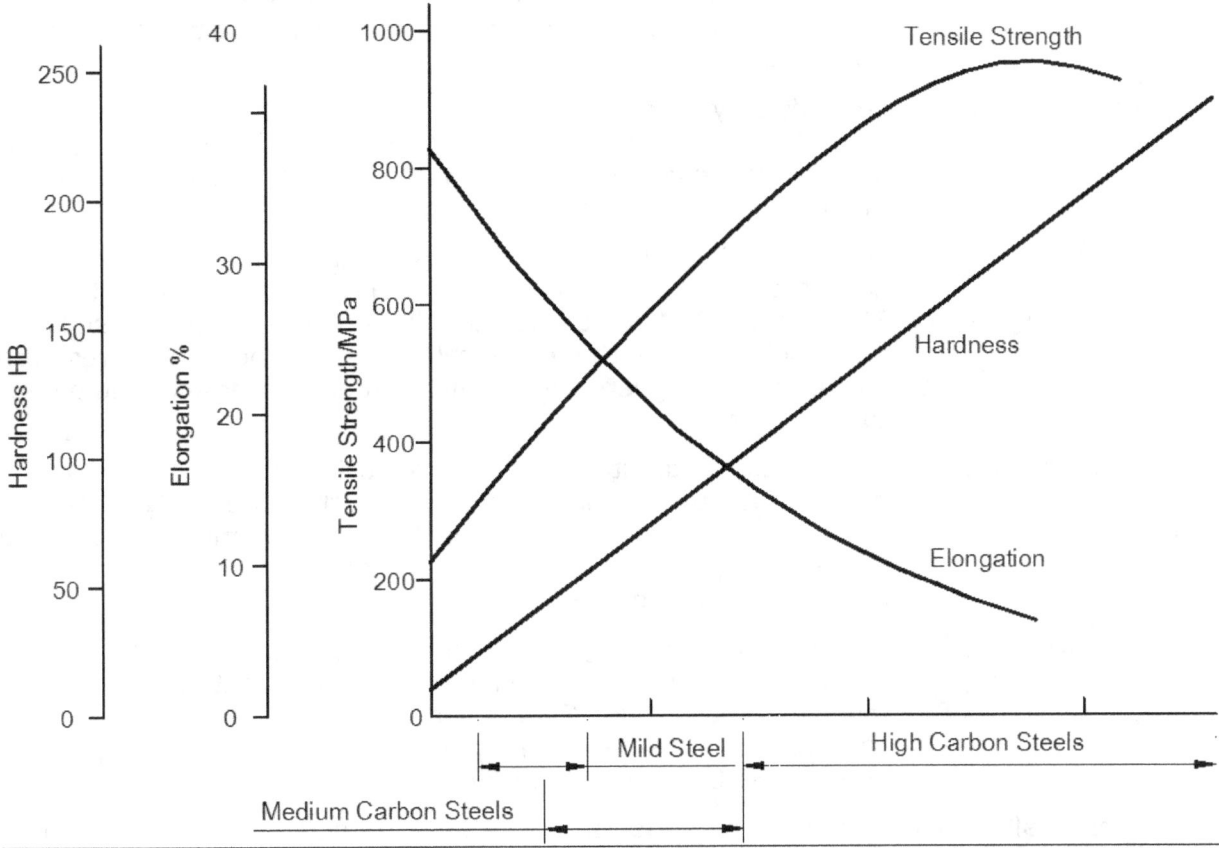

Figure 4.9 – Properties of Carbon Steels

Mild steel is general purpose steel and is used where hardness and tensile strength are not the most important requirements but ductility is often needed. Typical applications are sections for use as joists in buildings, bodywork for cars and ships and screws, nails and wire. Medium carbon steel is used for agricultural tools, fasteners, dynamo and motor shafts, crankshafts, connecting rods and gears. With such steels the lower ductility puts a limit on the types of processes that can be used. Medium carbon steels are capable of being quenched and tempered to develop reasonable toughness with strength. High-carbon steel is used for withstanding wear, where hardness is a more necessary requirement than ductility. It is used for machine tools, saws, hammers, cold chisels, punches, axes, dies, taps, drills and razors. The main use of high carbon steel is mainly as a tool steel. High carbon steels are usually quenched and tempered at about 250°C to develop their high strength with some slight ductility.

Example

An axe head may be made of high-carbon steel. Why use high carbon rather than mild steel?

High-carbon steel is a harder, stronger material than mild steel. The higher ductility of mild steel is not required in this situation.

Non-Ferrous Alloys:

The term non-ferrous alloy is used for all alloys where iron is not the main constituent, e.g. alloys of aluminium, of copper, of magnesium, etc. The following are some of the general properties and uses of non-ferrous alloys in common use in engineering.

Aluminium alloy — Low density, good electrical and thermal conductivity, high corrosion resistance. Tensile strengths of the order of 150 to 400 MPa, tensile modulus about 70 GPa. Used for metal boxes, cooking utensils, aircraft bodywork and parts.

Copper alloy — Good electrical and thermal conductivity, high corrosion resistance. Tensile strengths of the order of 180 to 300 MPa, tensile modulus about 20 to 28 GPa. Used for pump and valve parts, coins, instrument parts, springs, screws.

Magnesium alloy — Low density, good electrical and thermal conductivity. Tensile strengths of the order of 250 MPa and tensile modulus about 40 GPa. Used as castings and forgings in the aircraft industry where weight is an important consideration.

Nickel alloy — Good electrical and thermal conductivity, high corrosion resistance, can be used at high temperatures. Tensile strengths between about 350 and 1400 MPa, tensile modulus about 220 GPa. Used for pipes and containers in the chemical industry where high resistance to corrosive atmospheres is required, food processing equipment, gas turbine parts.

Titanium alloy — Low density, high strength, high corrosion resistance, can be used at high temperatures. Tensile strengths of the order of 1000 MPa, tensile modulus about 110 GPa. Used in aircraft for compressor discs, blades and casings, in chemical plant where high resistance to corrosive atmospheres is required.

Zinc alloy — Low melting points, good electrical and thermal conductivities, high corrosion resistance. Tensile strengths about 300 MPa, tensile modulus about 100 GPa. Used for car door handles, toys, car carburettor bodies - components that in general are produced by pouring the liquid metal into dies.

Copper alloy is an example of a non-ferrous alloy, consider. Pure copper is a soft material with low tensile strength. For many engineering purposes it is alloyed with other metals; the exception is where high electrical conductivity is required. Pure copper has a better conductivity than the alloys. The following indicate the names given to the various types of copper alloys:

Copper with zinc	Brasses
Copper with tin	Bronzes
Copper with tin and phosphorus	Phosphor bronzes
Copper with tin and zinc	Gun metals
Copper with aluminium	Aluminium bronzes
Copper with nickel	Cupro-nickels
Copper with zinc and nickel	Nickel silvers
Copper and silicon	Silicon bronze
Copper and beryllium	Beryllium bronze

Figure 4.10 shows how the percentage of zinc included with brasses affects the mechanical properties. Brasses with between 5% and 20% zinc are called gilding metals and are used for architectural and decorative items to give a gilded or golden colour. Cartridge brass is copper with 30% zinc; one of its main uses is for cartridge cases, items which require high ductility for the deep drawing process used to make them. The term basis brass is use for copper with 37% zinc and is a good alloy for general used with cold working processes and is used for fasteners and electrical connectors. It does not have the high ductility of those brasses with less zinc. Copper with 40% zinc is called Muntz metal.

The changes in the properties of brasses when the amount of zinc is altered arise from changes in the structure. Brasses with between 0% and 35% zinc form one type of structure, termed alpha, and between 5% and 45% there is a mixture of this alpha structure and another structure termed beta; it is this change in structure, i.e. the way the atoms of copper and zinc are packed together, that is responsible for the abrupt changes in properties of brass at 35% zinc.

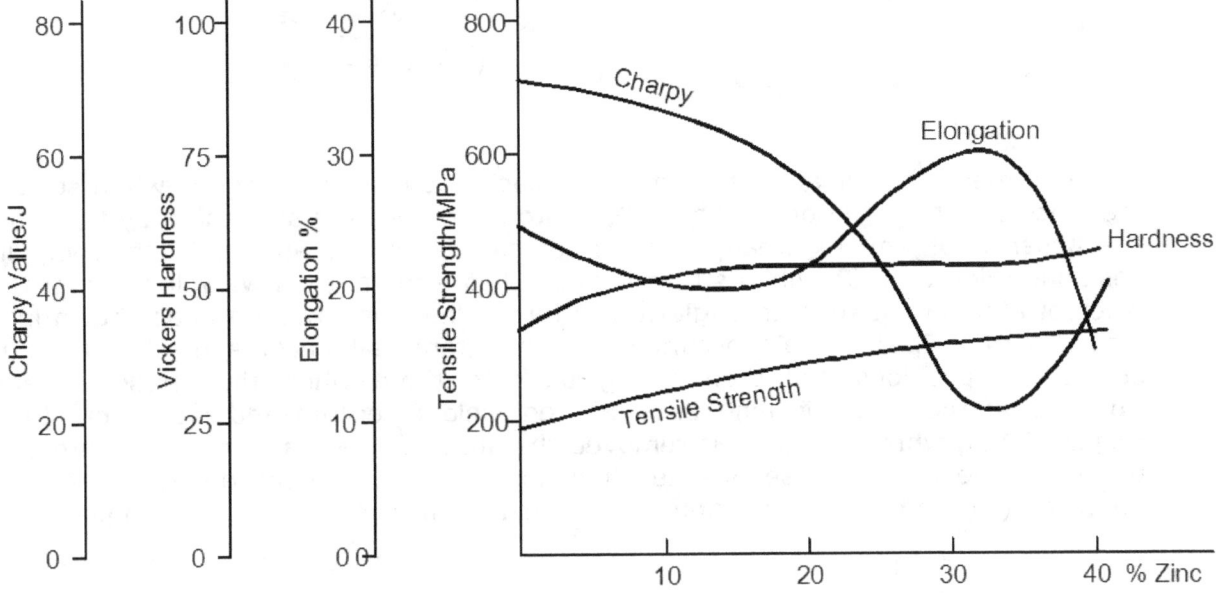

Figure 4.10 – Properties of Brasses

Stretching Metals:

A simple way we can think of the atoms in a metal is as though they were an array of spheres tethered to each other by springs, as illustrated in Figure 4.11. When forces are applied to stretch the material then the springs are stretched and exert an attractive force pulling the atoms back to their original positions. When forces are applied to compress the material then the springs are compressed and exert repulsive forces which push the atoms back to their original positions (Figure 4.12). We thus have a model for interatomic forces.

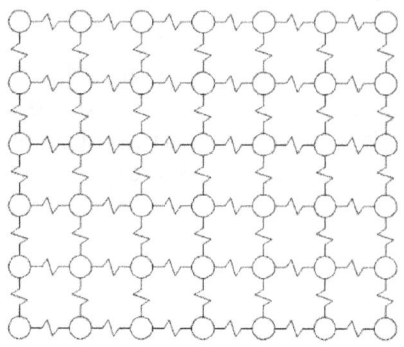
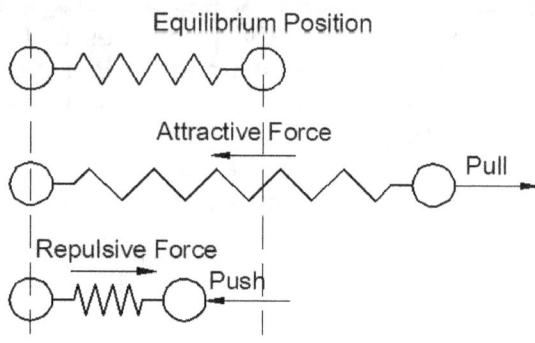

Figure 4.11   Figure 4.12

A simple theory to explain the elastic and plastic behaviour of metals when stretched is the block slip theory. Consider a block of atoms in the form suggested by the model. In the absence of any externally applied forces the atoms are all in their equilibrium positions (Figure 4.13); when stress is applied to a metal then we will consider that the block of atoms is at such an angle to the forces that the situation is as shown in Figure Figure 4.14. Elastic strain occurs when the atoms all become displaced from their equilibrium positions and then spring back to them when the stress is removed. However, if the stress is high enough then yielding occurs and blocks of atoms slip (Figure 4.15); when the stress is removed the atoms spring back to equilibrium positions but for some atoms these are new positions and permanent deformation has been produced (Figure 4.16). The plane along which atoms slip is called the slip plane.

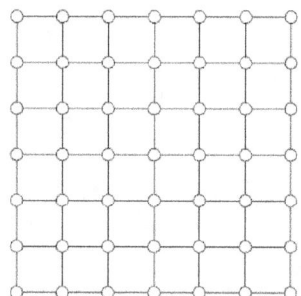
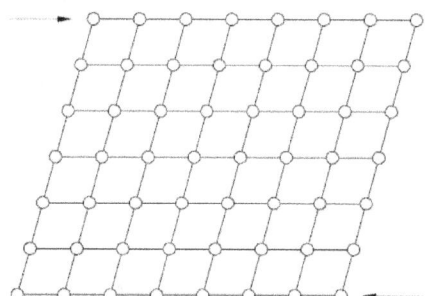

Figure 4.13 – No Stress Applied   Figure 4.14 – Stress Applied and Elastic Strain Produced

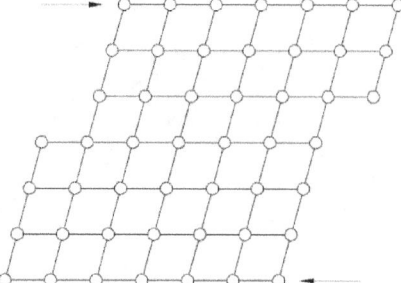
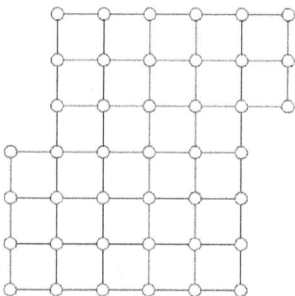

Figure 4.15 – Stress Applied and Yielding Occurring   Figure 4.16 – Stress Removed and Permanent Deformation Apparent

In terms of our model of a crystal as a pile of stacked spheres, slip has occurred when an entire row of spheres is pushed sufficiently for all to move along one position, as demonstrated in Figure 4.17 and Figure 4.18. On this model of slip, slip only transpires within an orderly arrangement of spheres, i.e. within a grain. Slip planes cannot thus cross over from one grain to another, the disorderly arrangement at the grain boundary does not allow it. Slip will only occur in those grains which have atomic planes at suitable angles to the applied forces; therefore a metal having big grains can result in more slippage than one having a larger number of small grains.

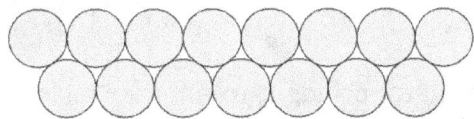
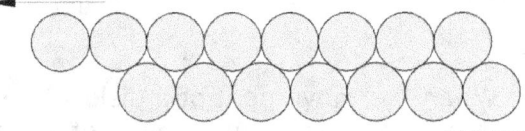

Figure 4.17 – Initial State

Figure 4.18 – After Slip

A simple model to consider is soldiers on parade in orderly ranks, i.e. all the soldiers in one large grain; for the movement of one soldier in the back rank to step forward then all the soldiers in that line step forward, there is slip and a large amount of movement. The analogy with the small grain metal is of a football crowd; when one person moves then there might be some local slip as other people move but there is no overall movement of the crowd. Thus, with metals, the bigger the grains, the greater the amount of plastic deformation that might be expected. A fine-grain structure will have less slippage and so show less plastic deformation, i.e. be less ductile. A brittle material is consequently one in which each slip process is confined to a short run in the metal and not allowed to spread. A ductile material is one in which the slip process is not confined to a short run in the metal and does spread over a large part.

While there can be considered to be many planes of atoms in a crystal, slip is found to occur only between the planes with the closest packing of atoms; this is because the atoms are close enough to more easily permit changes of positions than when further apart. Figure 4.19 illustrates this concept of close-packed planes, the lines indicating the planes with the highest density of atoms per unit length. If other lines are drawn through atoms less atoms per unit length will result. The number of such high density planes along which slip can occur depends on the form of structure of the crystal.

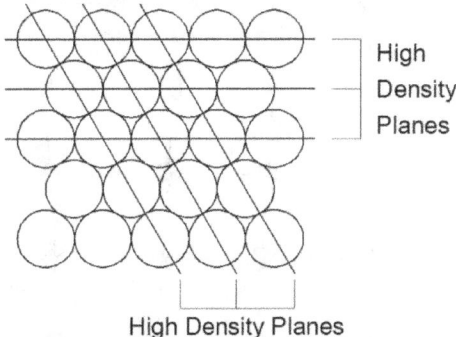

Figure 4.19 – High Density Planes

The body-centred cubic structure has many such slip planes, the face-centred cubic less and the hexagonal close-packed structure even less; thus metals which have a hexagonal close-packed structure tend to be harder and more brittle than those with the face centred cubic structure, while the body-centred cubic structure is likely to be the least hard and most ductile metal.

The above is just a simple model of what happens when metals are stretched. The model needs modification to do more than give a simple idea of what happens; an assumption has been made in the above model that the arrangement of atoms within a grain is perfectly orderly. In reality this is not the case and there are some atoms in the wrong places, the term dislocations being used; therefore, within a grain, the situation shown in Figure 4.20 might occur. When stress is applied the dislocation moves through the array of atoms, as illustrated in Figure 4.21, Figure 4.22 and Figure 4.23 so the slip takes place atom by atom rather than the wholesale movement of one plane of atoms past another.

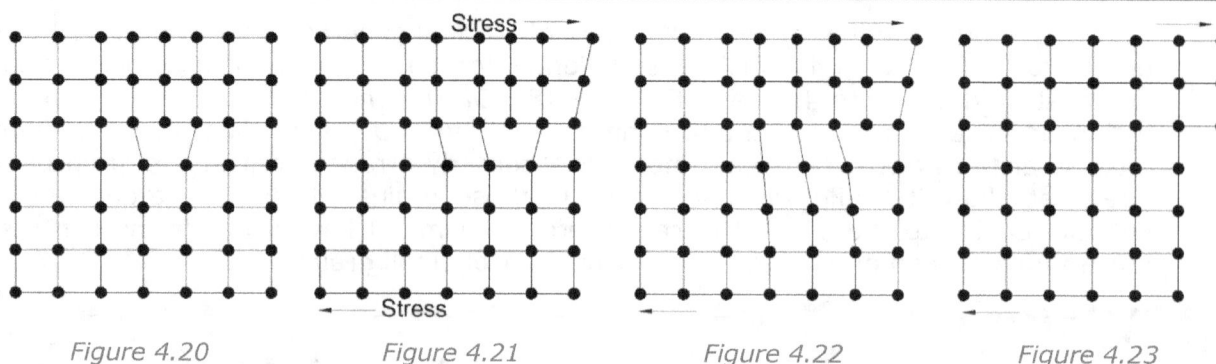

Figure 4.20   *Figure 4.21*   *Figure 4.22*   *Figure 4.23*

When the movement of a dislocation through a metal brings it to another dislocation then they can either cancel each other out or hinder further movement. Figure 4.24 shows what can happen when two dislocations come close to each other; each dislocation has the atoms on one side of the slip plane in compression and on the other side in tension. When two compression regions come close together the forces between the atoms result in the dislocations repelling each other. In general, the more dislocations a metal has, the more the dislocations get in the way of each other and so the more difficult it is for the dislocations to move through the metal and hence slip to occur; consequently the greater the number of dislocations, the greater the stress needed to produce yielding.

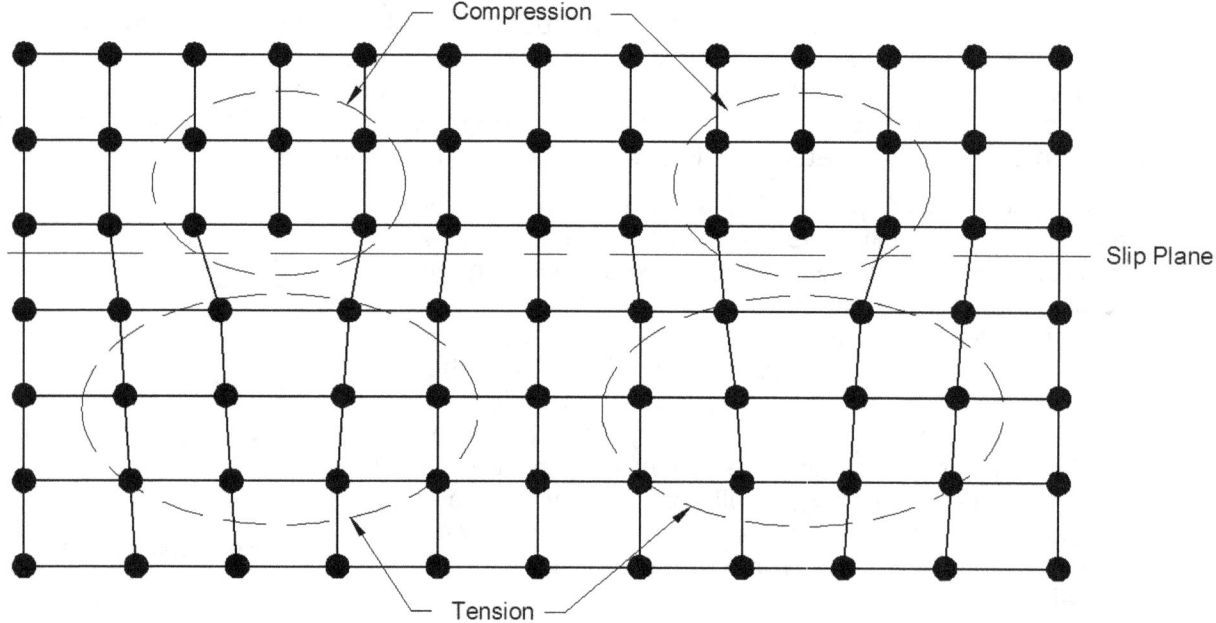

Figure 4.24 – Dislocations Repelling Each Other

Dislocations are produced as a result of missing atoms, atoms being displaced from their correct positions, and foreign atoms being present and distorting the orderly packing of atoms. Cold working distorts grains, resulting in an increase in dislocations because it displaces atoms from their correct positions. The foreign atoms may be present as a result of a deliberate alloying process, therefore alloying, in increasing the number of dislocations and making it more difficult for dislocations to move through the material, increases the yield stress. Another method of increasing the yield stress is to cause small particles to precipitate out from an alloy. Such a process is called precipitation hardening.

Cold Working:

Cold working of metal is done by subjecting metal to enough mechanical stress to cause plastic deformation, a permanent change in the metal's crystalline structure. Cold working is named because the work is done at temperatures below the metal's recrystallization point and alters the metal's structure through mechanical stress rather than heat. The technique increases a metal's strength and hardness while reducing its ductility. A number of different processes are used in the modern metalworking industries that are applied to materials such as steel, aluminium and copper.

Cold working of metal strengthens the material through a process called work hardening or strain hardening; when the mechanical stress on a metal becomes high enough, it causes permanent crystallographic defects, called dislocations, in the crystalline structure of the metal's atoms. As the number of dislocations increases, it becomes more difficult for new ones to form or for the existing defects to move through the crystal structure, making the metal become more resistant to further deformation increasing the yield strength and allowing it to withstand greater stress, but it also means that the metal becomes less ductile and if the metal is subjected to too much stress, it will fracture rather than bend.

Cold working is often more cost effective than working metal through heat treatment, especially for large-volume production, because it produces comparable improvements in strength while using materials more efficiently and requiring less finishing. The high initial capital cost of this process, however, makes it less cost effective than heat treatment at smaller scales. The lower ductility of cold worked metal also makes it inferior in some applications. Its higher resistance to deformation makes it less able to give way to forces the metal is not strong enough to resist, and so if the metal is subjected to too much stress, it can fracture rather than bend. Some metal production uses both methods at different points in the production process to impart the desired qualities in the metal.

There are a number of different methods that can be used for cold working; the most common type is cold rolling, in which the metal being worked is squeezed through narrow gaps between rotating metal rolls. The movement of the rolls compresses the material, causing deformation as moves it through the gap. Another method is cold forging, in which metal is shaped by forcing it into a die with a press or hammer.

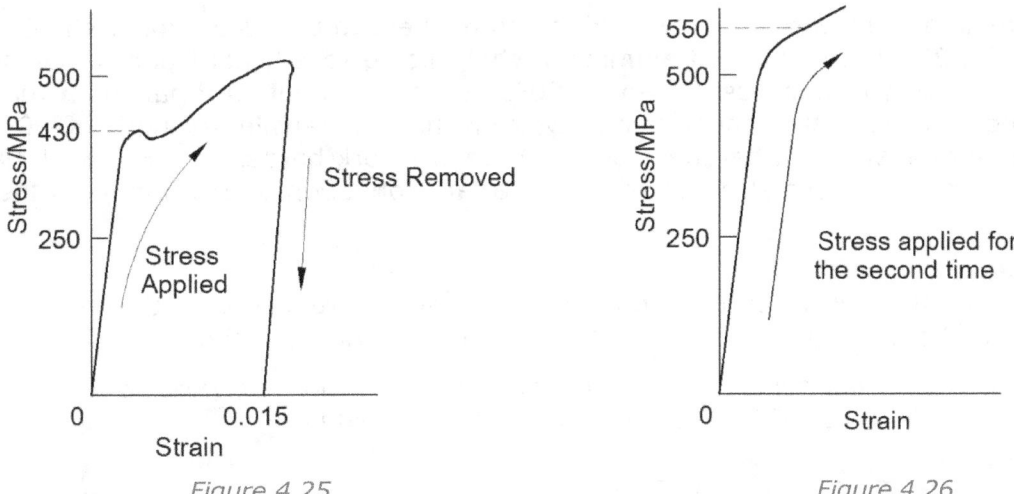

Figure 4.25   Figure 4.26

As an example of cold working, a tensile test is performed on a carbon steel test piece with the material showing a yield stress of 430 MPa. If the stress is continued beyond this point but the stress released before the material breaks then a permanent deformation will be found to have occurred.

Figure 4.25 illustrates this sequence of events and indicates a permanent deformation of strain of 0.015. Now suppose the material is again stretched. This time the yield stress is not 430 MPa but 550 MPa. The material has now a much higher yield stress (Figure 4.26); it is not only the yield stress which has increased, the tensile strength has increased, the percentage elongation has decreased, and the hardness has increased.

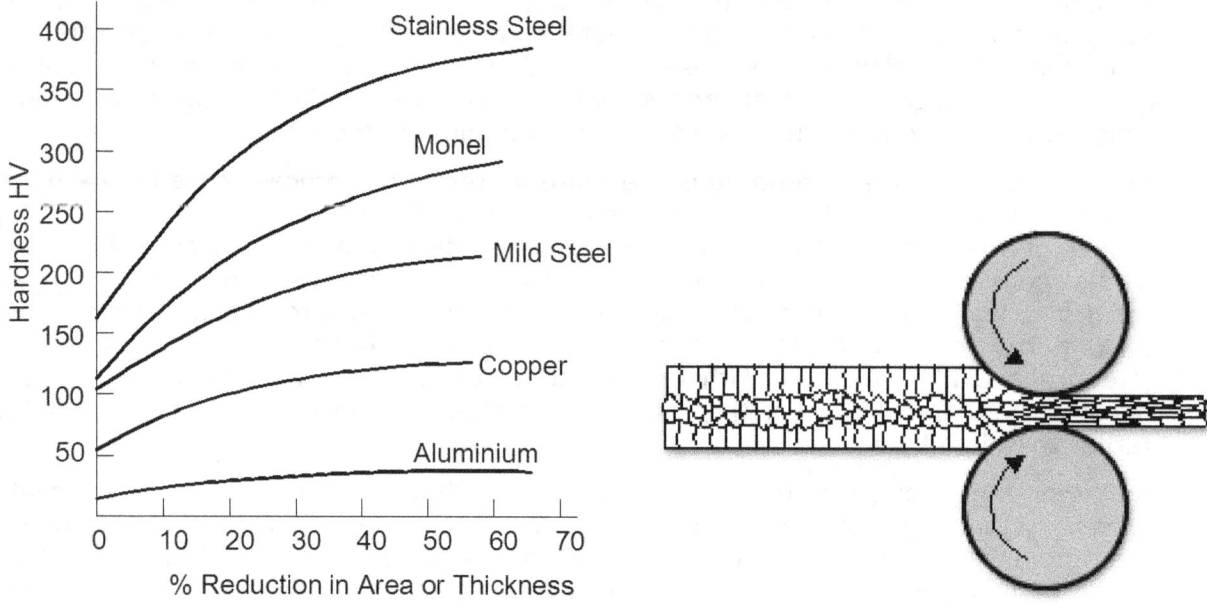

Figure 4.27

Figure 4.28 – Cold Rolling

The material is said to have been subject to cold working and the above are typical of the changes that occur. The term cold working is used when plastic deformation is produced at a temperature which is not high enough to produce changes. The term work hardening is often used since the cold working has resulted in the material becoming harder. Figure 4.27 shows the effect of cold working on the hardness of typical materials. The more a material is worked, the harder it becomes. A stage is reached, however, when the hardness is at a maximum and further deformation is not possible as the material is too brittle. For example, with the aluminium referred to in the Figure this is when the aluminium has been reduced in thickness by about 60%. The material is then said to be fully work hardened.

An example of cold working is the cold rolling of sheet to produce thinner sheet as shown in Figure 4.28. With annealed aluminium sheet being rolled, full work hardening occurs with a reduction in thickness of about 60% and so is about the maximum that can be produced. The annealed sheet might typically have a tensile strength of 90 MPa, an elongation of 40% and a hardness of 20 HV before work hardening. Full work hardening might result in a tensile strength of 150 MPa, an elongation of 3% and a hardness of 40 HV.

Example
Using Figure 4.27, what is the approximate percentage reduction in thickness of a sheet of mild steel that is possible before it becomes fully work hardened?

The mild steel would appear to have reached its maximum hardness with a reduction in thickness of about 50 - 60% and so be fully work hardened.

<u>**Heat Treating Cold-Worked Metals:**</u>
Cold working of metals results in changes in mechanical properties with yield stress, tensile strength, and hardness increasing and percentage elongation decreasing. These changes can be reversed by suitable heat treatment.

Cold-worked metals generally have deformed grains, with a high density of dislocations within the grains. Such dislocations lead to internal stresses within grains. When such a metal is heated then there is some slight rearrangement of atoms within the grains and a consequent reduction in internal stresses; this process is known as recovery.

When a cold-worked metal is heated above about $0.3T_m$, where T_m is the melting point of the material on the kelvin scale of temperature, then there is a marked reduction in tensile strength, hardness and an increase in percentage elongation. Figure 4.18 illustrates this for cold-worked copper; what is happening is that the material is beginning to recrystallize with new grains starting to grow. The temperature at which recrystallization starts is called the recrystallization temperature. For pure metals it tends to be about 0.3 to $0.5T_m$; therefore, aluminium which has a melting point of 933 K has a recrystallization temperature of 423 K, about 0.45 Iron with a melting point of 1356 K has a recrystallization temperature of 473 K, about 0.35 T_m.

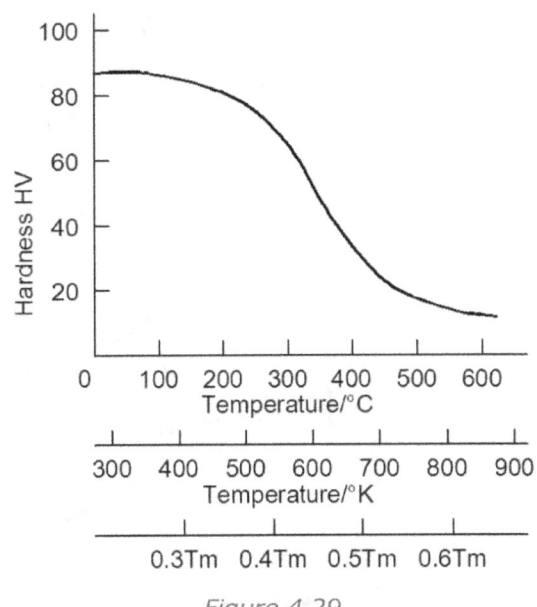

Figure 4.29

The term annealing is used for the heat treatment process which involves heating the material to above the recrystallization temperature and obtaining more ductile properties.

As the temperature is further increased so the crystals start to grow until they have completely replaced the original distorted cold worked structure. Figure 4.30 illustrates the sequence of events.

Example

A manufacturer of copper sheet receives the copper as much thicker plate, with already some amount of cold working. It is proposed to produce the sheet by cold rolling in a number of stages, the stages being separated by annealing. Why is the sheet production in a number of stages?

If copper is cold worked to about a 60% reduction in thickness it becomes brittle and tends to break with further working; also it becomes fairly hard and difficult to roll. By following the rolling by annealing the material is made more ductile again and rolling can continue without difficulty.

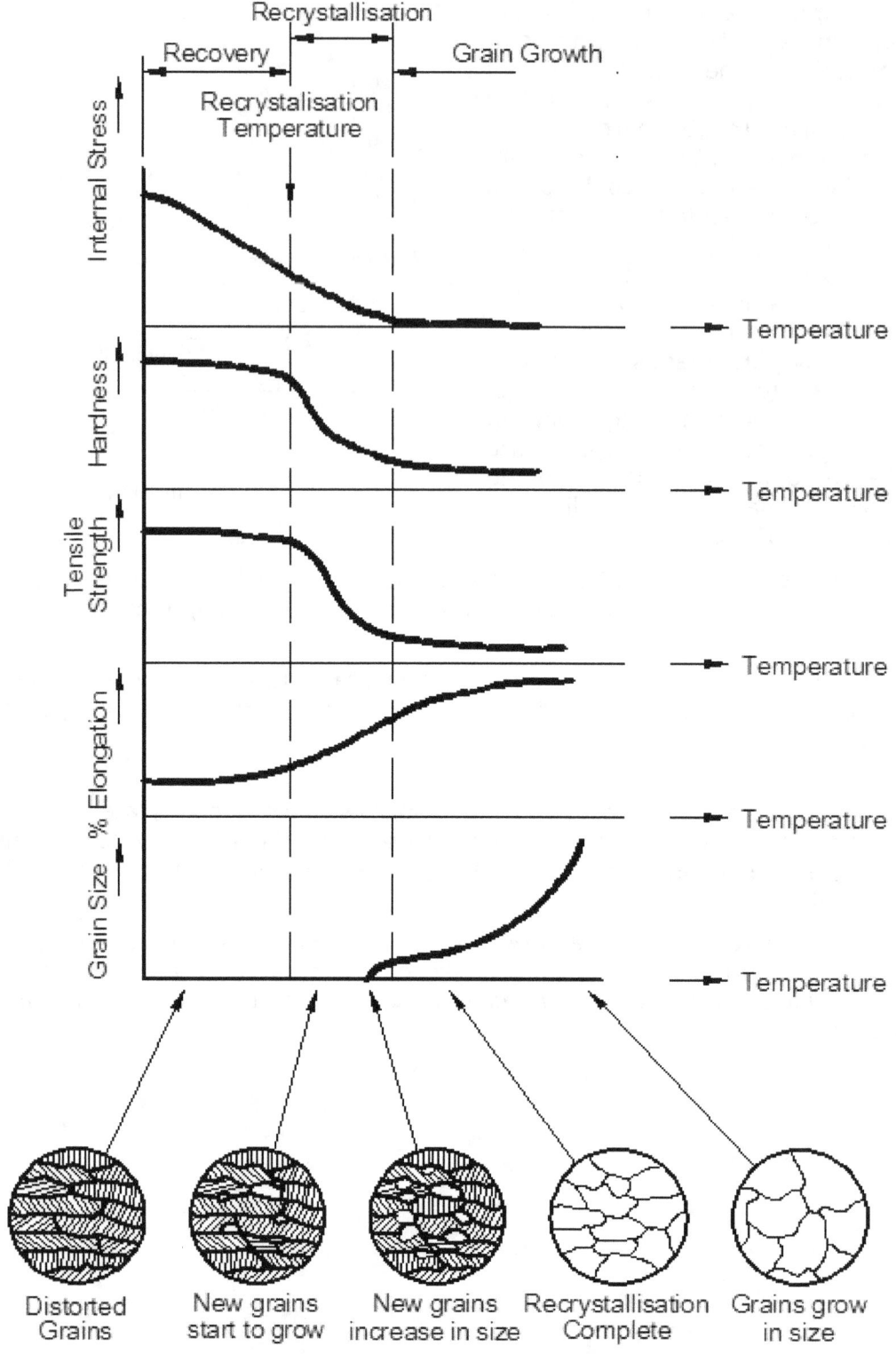

Figure 4.30 – The Effect of Heat Treatment on Cold-Worked Metals

Hot Working:

Hot working contrasts with cold working and involves deforming a material at a temperature greater than the recrystallization temperature then, as soon as a grain becomes deformed it recrystallizes. No hardening occurs and the working can be continued without any difficulty and no interruption of working is needed to anneal the material, as is the case with cold working.

A disadvantage of hot working is that oxidization of the metal surfaces occurs. Cold working does not have this problem; another disadvantage is that the material will have comparatively low values of hardness and tensile strength, with high percentage elongation. A combination of hot and cold working is thus often used in a particular shaping process. The first operation, involving large amounts of plastic deformation, is carried out by hot working. After cleaning the surfaces of the metal, it is then cold worked to increase the strength and hardness and give a good surface finish.

Example

Lead has a melting point of 327°C. Will the product be work hardened if it is made by extruding at room temperature? Extrusion is a process rather similar to the squeezing of toothpaste out of a tube, the metal being squeezed out through a nozzle and taking the shape dictated by that of the nozzle.

The melting point of lead is about 600 K. This would mean that the extrusion at about 300 K is at about 0.5 T_m. In other words, the extrusion is taking place at about the recrystallization temperature. The process is likely to be just about a hot working process and so there would be no work hardening.

The Structure of Polymers:

Everyday household utensils which have polymer molecules as their basis are the plastic washing-up bowl, the plastic measuring rule, the plastic cup. A polymer molecule in a plastic may have thousands of atoms all joined together in a long chain. The backbones of these long molecules are chains of carbon atoms. Carbon atoms are able to form strong bonds with themselves and produce long chains to which other atoms can become attached.

The term polymer is used to indicate that a compound consists of many repeated structural units. The prefix poly means many, each structural unit in the compound is called a monomer. For many plastics the monomer can be deduced by deleting the prefix poly from its name consequently the plastic called polyethylene is a polymer which has as its monomer ethylene. Figure 4.31 shows the monomer and the resulting polymer.

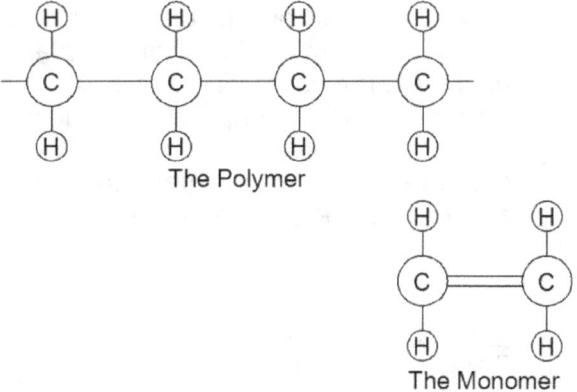

Figure 4.31 - Polythene

Figure 4.32, Figure 4.33 and Figure 4.34 shows the basic forms that can be adopted by the molecular chains; these forms can be described as linear, branched and cross-linked chains. The linear chains have no side branches or links with other chains and can thus move readily past each other. If, however, the chains have side branches there is a reduction in the ease with which chains can move past each other and so the material is more rigid. If there are crosslinks a much more rigid material is produced in that the chains cannot slide past each other at all.

Figure 4.32
Linear Polymer Chain

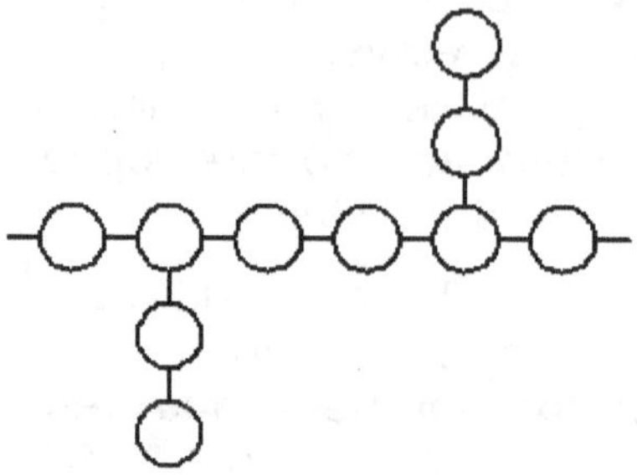

Figure 4.33
Broached Polymer Chain

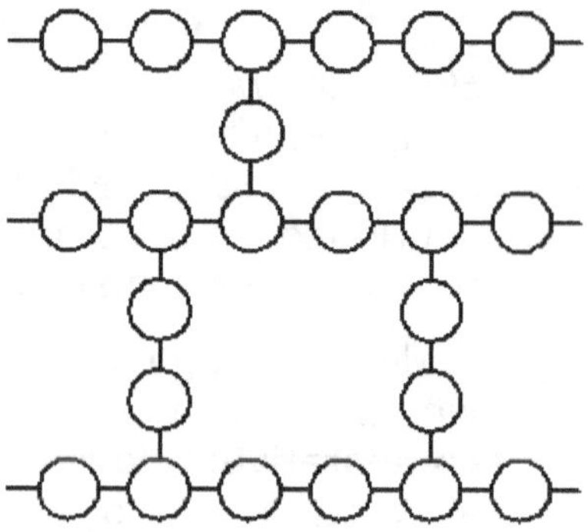

Figure 4.34
Cross-Linked Polymer

Polymers can be classified as thermoplastics, thermosets or elastomers. A simple method by which thermoplastics and thermosets can be distinguished is when heat is applied. With a thermoplastic the material softens with removal of the heat resulting in hardening, whereas, with a thermoset, heat causes the material to char and decompose with no softening. An elastomer is a polymer which by its structure allows considerable extensions which are reversible; thermoplastics have linear or branched chains for their structure. Thermosets have a cross-linked structure and elastomers are chains with some degree of cross-linking.

Additives:

The term plastic is commonly used to describe materials based on polymers. Such materials, however, invariably contain other substances which are added to the polymers to give the required properties. Since some polymers are damaged by ultraviolet radiation, protracted exposure to the sun can lead to a deterioration of mechanical properties. An ultraviolet absorber is thus often added to the polymer, such an additive being called a stabilizer. Carbon black is often used for this purpose. Plasticizers are added to the polymer to make it more flexible. In one form this may be liquid which is dispersed throughout the solid, filling the space between the polymer chains and acting like a lubricant and permitting the chains to slide past each other more easily. Flame retardants may be added to improve fire-resistant properties and pigments and dyes to give colour to the material. The properties and cost of a plastic can be markedly affected by the addition of substances termed fillers. Since fillers are generally cheaper than the polymer, the overall cost of the plastic is reduced. Often up to 8 0 % of a plastic may be filler; examples of fillers are glass fibres to increase the tensile strength and impact strength, mica to improve electrical resistance, graphite to reduce fiiction and wood flour to increase tensile strength. One form of additive used is a gas to give foamed or expanded plastics. Expanded polystyrene is used as a lightweight packaging material and foamed polyurethane as a filling for upholstery and sponges.

Thermoplastics:

Plastics are considered as one of the most unfriendly materials that humans have produced. Plastics are not bio-degradable and bulky to collect and store.

Thermoplastics consist of polymers with long-chain molecules which are either linear chains or long chains with small branches. Linear chains have no side branches or cross-links with other chains; because of this they can easily move past each other. If, however, the chain has branches then there is a reduction in the ease with which chains can be made to move past each other; this shows itself in the material being more rigid, i.e. less strain produced for a given stress.

A crystalline structure is one in which there is an orderly arrangement of particles and a structure in which the arrangement is completely random is said to be amorphous. Many polymers are amorphous with the polymer chains being completely randomly arranged in the material (Figure 4.35). Linear polymer molecules can, however, assume an arrangement which is orderly. Figure 4.36 shows the type of arrangement of chains that can occur, the linear chains folding backwards and forwards on themselves. The arrangement is said to be crystalline. The tendency of a polymer to crystallize is determined by the form of the polymer chains. Linear polymers can crystallize to quite an extent, but complete crystallization is not obtained in that there are invariably some regions of disorder. Polymers with side branches show less tendency to crystallize since the branches get in the way of the orderly arrangement. The greater the crystallinity of a polymer, the closer the polymer chains can be packed and so the greater the density.

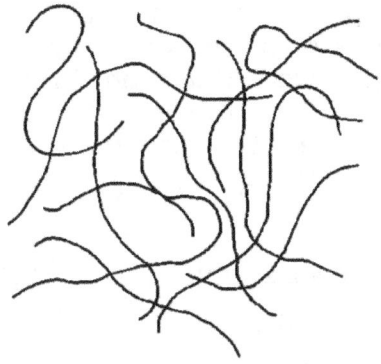

Figure 4.35
Linear Amorphous Polymer

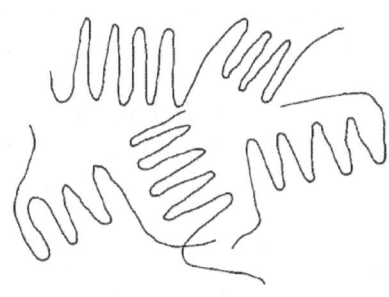

Figure 4.36
Folded Linear Polymer

Figure 4.37 shows the typical form of a stress-strain graph for a crystalline polymer. When stress is applied the first thing that begins to happen is that some movement of folded chains past each other occurs. When point A is reached the polymer chains start to unfold to give a material with the chains all lying along the direction of the forces stretching the material. The material shows this by starting to exhibit necking (Figure 4.38). As the stress is further increased, the necking spreads along the material with more and more chains unfolding. Eventually, when the entire material is at the necked stage all the chains have lined up. The material in this state behaves differently from earlier in the stress-strain graph, the material being said to be cold drawn; the sequence of events tends to occur only if the material is stretched slowly and sufficient time elapses for the molecular chains to unfold. If a high strain rate is used the material is likely to break without the chains all becoming lined up; the plastic used for making polythene bags is a crystalline polymer. Try cutting a strip of polythene from such a bag and pulling it between your hands and see the necking develop with low rates of strain.

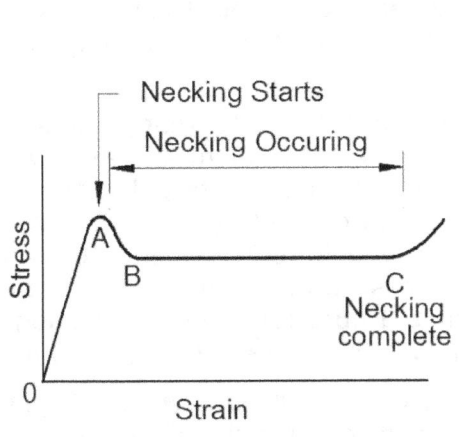

Figure 4.37

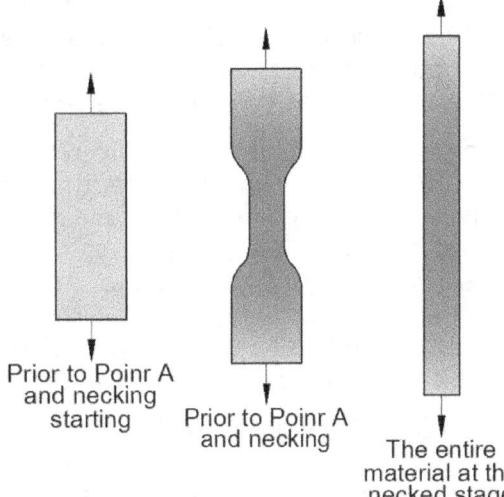

Figure 4.38

Examples of Thermoplastics:

Examples of thermoplastics are polyethylene (polythene), polypropylene, polyvinyl chloride (PVC), polystyrene, acrylonitrile-butadiene-styrene terpolymer (ABS) and polyamides (nylons). The following is a brief discussion of the structure of these thermoplastics and their properties.

Polyethylene

Polyethylene is a type of polymer that is thermoplastic, meaning that it can be melted to a liquid and remoulded as it returns to a solid state. It is chemically synthesized from ethylene, a compound that's usually made from petroleum or natural gas. Other non-official names for this compound include polythene or polyethylyne and abbreviated as PE. It is used in making other plastic compounds much often than it's used in its pure form; though it has a wide variety of uses, it can be harmful to humans and to the environment. Polythene is available in either high-density or low-density; the following table gives a comparison of the properties of the two forms.

Properties of Polyethylene

Property	Low Density	High Density
Crystallinity	60%	95%
Density (10^3 kg/m3)	0.92	0.95
Melting Point (°C)	115	138
Tensile Strength (MPa)	8 to 16	220to 38
Tensile Modulus	0.4 to 0.3	0.4 to 1.3
% Elongation	100 to 600	50 to 800
Max. Service Temperature (°C)	85	125

Of all the plastics produced for industrial and commercial products, polyethylene is the most common. As an example, 1.3 million metric tonnes of it is produced each year in Australia alone with over 69% originating from packaging and materials handling. Over five times as much PE is manufactured each year than a closely-related compound, polypropylene (PP). The largest use for these polymers is in packaging materials, like films and foam; and for bottles and other containers that can be used in food, medical, and other consumer industries.

Polypropylene

Polypropylene, is a plastic polymer with the chemical formula C_3H_6. It is used in many different settings, both in industry and in consumer goods, and it can be used both as a structural plastic and as a fibre; this plastic is often used for food containers, particularly those that need to be dishwasher safe.

The melting point of polypropylene is very high compared to many other plastics, at 320°F (160°C), which means that the hot water used when washing dishes will not cause dishware made from this plastic to warp and contrasts with polyethylene, another popular plastic for containers, which has a much lower melting point. Polypropylene is also very easy to add dyes to, and it is often used as a fibre in carpeting that needs to be rugged and durable, such as that for use around swimming pools or on miniature golf courses. Unlike nylon, which is also often used as a fibre for rugged carpeting, it doesn't soak up water, making it ideal for uses where it will be constantly subject to moisture.

Research is ongoing with polypropylene, as makers experiment with different methods for synthesizing it; some of these experiments yield the promise of exciting new types of plastic, with new consistencies and a different feel from the fairly rigid version that most people are used to. These new elastic versions are very rubbery, making them even more resistant to shattering and opening up many different uses for an already pervasive plastic.

Polypropylene is not as sturdy as polyethylene, but it has benefits that make it the better choice in some situations. One of these situations is creating hinges from a plastic, such as a plastic lid on a travel mug. Over time, plastics wear out from the repetitive stress of being opened and shut, and eventually will break. Polypropylene is very resistant to this sort of stress, and it is the plastic most often used for lids and caps that require a hinging mechanism.

Like many plastics, polypropylene has virtually endless uses, and its development has not slowed since its discovery. Whether used for industrial moulds, rugged currency, car parts, or storage containers, it is one of a handful of materials the world is literally built around.

Properties of Polypropylene

Property	
Crystallinity	60%
Density (10^3 kg/m^3)	0.90
Melting Point (°C)	176
Tensile Strength (MPa)	30 to 40
Tensile Modulus	1.1 to 1.6
% Elongation	50 to 600
Max. Service Temperature (°C)	150

Polyvinyl chloride (PVC)

Polyvinyl chloride, better known as PVC or vinyl, is an inexpensive plastic so versatile it has become completely pervasive in modern society. The list of products made from polyvinyl chloride is exhaustive, ranging from phonograph records to drainage and potable piping, water bottles, cling film, credit cards and toys. More uses include window frames, rain gutters, wall panelling, doors, wallpapers, flooring, garden furniture, binders and even pens. Even imitation leather is a product of polyvinyl chloride. In fact, it's hard to turn anywhere without seeing some form of this plastic.

In 1913, polyvinyl chloride became the first synthetic product ever patented; however, its diversity and ubiquitousness is now in question, as it comes from a highly toxic production industry and potentially remains an environmental threat throughout all phases of its life. In addition to the toxic chemical processing required to make PVC, mounting research indicates a tendency for some PVC products to leech harmful chemicals, with a possible link to health risks and environmental contamination.

Additionally, polyvinyl chloride is not biodegradable, a fact that manufacturers promote as a plus, while environmentalists count it among many of polyvinyl chloride's drawbacks. They point to the ever-growing massive amounts of discarded PVC products and shrinking landfills, and the potential for long-term leeching that could lead to ground water contamination. Polyvinyl chloride should not be burned, as it can release harmful gas, and recycling is difficult because of the diverse additives used in various products.

One of the by-products of the polyvinyl chloride manufacturing process is organochlorine. Though chlorine is found naturally in the environment in minerals such as salt, this type of chlorine is different. Highly reactive, its effect in concentrated form can be very destructive, as seen in other manufacturing industries. Some familiar forms of organochlorines include polychlorinated biphenyls (PCBs), banned in the 1970s; halon and CFCs, responsible for destroying the ozone; and DDT. Purportedly, the production of polyvinyl chloride results in the generation of more organochlorines than any other material.

Aside from the environment, human health is also a concern. Studies regarding initial outgassing of chemicals from polyvinyl chloride plastics like those used in shower curtains, flooring and vinyl car interiors are ongoing. Leeching of a softening chemical called DEHP (di-2-ethyl hexyl phthalate) in products like vinyl IV bags used in the neonatal wards of some hospitals has also been a concern. Alternate softening agents are reportedly under consideration by the PVC industry but require further testing.

Though polyvinyl chloride products have been used without apparent problems to human health for many years, the concern is that growing toxic waste created by the process, possible leeching, and PVC's non-biodegradable status will eventually and inevitably lead to problems that could be catastrophic. The conservative trend is headed towards environmentally friendly, biodegradable alternatives. Among others, these include wood, paper, copper, steel, and clay. Chlorine-free plastics, such as polyethylene (PE), polypropylene (PP) and polyisobutylene, may also be preferred over PVC, although most of these are not biodegradable.

Properties of PVC

Property	No Plasticizer	Low Plasticizer	High Plasticizer
Crystallinity	0%	0%	0%
Density (10^3 kg/m^3)	1.4	1.3	1.2
Tensile Strength (MPa)	52 to 58	28 to 42	14 to 21
Tensile Modulus	2.4 to 4.1		
% Elongation	2 to 40	200 to 250	350 to 450
Max. Service Temperature (°C)	70	60 to 100	60 to 100

Polystyrene

Polystyrene is a type of polymer with thermoplastic properties produced from the petroleum-derived monomer, styrene. In solid form, it is a colourless and rigid plastic, but it may also be returned to a liquid state by heating, and used again for moulding or extrusion; it is used to produce many products for industrial and consumer use and in its presence as a plastic in everyday life is second only to polyethylene.

The chemical structure of this material allows it to be classified as a liquid hydrocarbon, meaning that it is composed exclusively of hydrogen and carbon. Like its precursor, it's an aromatic hydrocarbon that participates in covalent bonding with every other carbon atom being attached to a phenol group. It is produced via free radical polymerization, which means that the reaction involves breaking the bonds between electrons and leaving them free to form new bonds. When burned, this material yields black carbon particles, or soot. When completely oxidized, only carbon dioxide and water vapour remain.

There are several different types that are produced. Extruded polystyrene is considered to have as much tensile strength as unalloyed aluminium but it is lighter and more elastic. This is the material used to make a variety of moulded products, ranging from plastic tableware to CD cases and model cars; it is also used to produce medical and pharmaceutical supplies.

Extruded polystyrene foam, commonly known as Styrofoam is a type of insulation with versatile applications, such as the manufacture of surfboards. In fact, its buoyancy prompted marine life-saving industry to adopt its use in life rafts. This type may also be used in building materials construction. For example, it may serve as a layer of insulation under a concrete slab to assist in the distribution of loads and save time digging trenches.

This strong but lightweight material is also used for crafts, and it is usually sold in sheets. It is typically made up of three layers, with polystyrene at the core sandwiched by paper on either side. The sheets are frequently used as backing to mount artwork or photography, or to construct architectural models. This foam is also familiar to those who work with floral crafts. In fact, the characteristic "crunch" sound made when it's cut is well-known to florists.

Expanded polystyrene foam is actually made of beads of the material. It is used to make beads for packaging, disposable coffee cups, and foam picnic coolers. Unlike extruded foam, which is blue in colour, expanded foam is typically white.

Properties of Polystyrene

Property	No Additives	Toughened
Crystallinity	0%	0%
Density (103 kg/m3)	1.1	1.1
Tensile Strength (MPa)	35 to 60	17 to 42
Tensile Modulus	2.5 to 4.1	1.8 to 3.1
% Elongation	1 to 3	8 to 50
Max. Service Temperature (°C)	65	75

Acrylonitrile-Butadiene-Styrene Terpolymer (ABS)

Acrylonitrile contributes with thermal and chemical resistance, and the rubberlike butadiene gives ductility and impact strength. Styrene gives the glossy surface and makes the material easily machinable and less expensive.

Generally, ABS has good impact strength also at low temperatures. It has satisfactory stiffness and dimensional stability, glossy surface and is easy to machine. If UV-stabilizers are added, ABS is suitable for outdoor applications. Acrylonitrile Butadiene Styrene was first discovered during World War II when its basis, SBR, was used for alternatives to rubber. Commercially acrylonitrile butadiene styrene polymers first became available in the early 1950s in an attempt to obtain the best properties of both polystyrene and styrene acrylonitrile.

Properties of ABS

Property	
Crystallinity	0%
Density (103 kg/m3)	1.1
Tensile Strength (MPa)	17 to 58
Tensile Modulus	1.4 to 3.1
% Elongation	10 to 140
Max. Service Temperature (°C)	110

Polyamides

The first polyamide plastic was prepared in 1934 although it was not until 1937 that production was started commercially. The new plastic was called nylon. The chief raw materials used in the production of nylon are benzene and butadiene ($CH_2=CH-CH=CH_2$). Various intermediates, particularly caprolactam, are produced before the final product is obtained.

Polyamides are chiefly used for making moulded articles (including household fixtures and fittings) and textiles such as garments, gears, bearings, bushes, housings for domestic and power tools, electric plugs and sockets.

Properties of Nylons

Property	Nylon 6	Nylon 6.6
Crystallinity	Can be varied from low to high percentages	
Density (10^3 kg/m³)	1.13	1.1
Melting Point	225	265
Tensile Strength (MPa)	75	80
Tensile Modulus	1.3 to 3.1	2.8 to 3.3
% Elongation	60 to 320	60 to 300
Max. Service Temperature (°C)	110	110

Thermosets:

Thermoplastics are materials that take a hardened form after they are heated and allowed to cool. When these materials are heated again, they generally convert to liquid and can be reformed. Thermosets also take a hardened form after they have been heated and allowed to cool. A major difference is that thermosets cannot be melted down and reformed. There are many types of thermosets, such as vulcanized rubber and epoxy resin.

Some materials, such as thermoplastics, can be found in forms that can be changed with heat. A simplified way of drawing a contrast is to view those items as many molecules that have been melted together but whose bonds can be released upon reheating. With thermosets, however, when the materials are heated, the molecules merge irreversibly. Reheating will not release the bonds. Instead, reheating is most likely to destroy the materials.

For this reason, thermoset materials are typically considered to be non-recyclables; for many people, this is a major disadvantage. There are many factors about thermosets, however, that can be seen as advantages including strength and durability.

Before thermosets are made, the materials are often in liquid form or another form that makes them susceptible to shaping. The process that those materials are put through to create the finished form is known as curing; there are several types of curing processes with each tending to produce different types of materials.

One curing process is vulcanization, which is used to make vulcanized rubber for products such as tires, bowling balls, and hoses. There are several methods of vulcanization, but overall they are all generally considered irreversible processes. The rubber produced tends to differ from natural rubber in several ways; it is less sticky, resistant to heat, and more capable of holding the desired shape.

Some thermosets are produced when an epoxide, such as epichlorohydrin, is mixed with a hardener, such as Bisphenol-A. Completion of such a polymerization process can result in epoxy resins; these materials are considered very versatile because most of their characteristics can be altered by making modifications during the process. Epoxy resins tend to have excellent chemical and heat resistance.

Epoxy resins are used in numerous industries. Artists use the material as a painting medium. In the electronics field, epoxy resin is used to produce circuit boards and transistors and is used as an adhesive.

Examples of Thermosets:

The following is a brief outline of the nature and properties of commonly used thermosets. Such materials are widely used with open-weave fabrics, such as glass-fibre fabric, to give composites.

Phenolics give highly cross-linked polymers. Phenol formaldehyde was the first synthetic plastic and is known as Bakelite. The polymer is opaque and initially light in colour; it does, however, darken with time and so is always mixed with dark pigments to give a dark-coloured material. It is supplied in the form of a moulding powder which includes the polymer, fillers and other additives such as pigments. When this moulding powder is heated in a mould the cross-linked polymer chain is produced. The fillers account for some 50-80% of the total weight of the moulding powder. Wood flour which is a very fine softwood sawdust, when used as a filler increases the impact strength of the plastic, asbestos fibres improve the heat properties, and mica the electrical resistivity. The following table shows some of the properties of this thermoset. Phenol formaldehyde mouldings are used for electrical plugs and sockets, switches, door knobs and handles, camera bodies and ash trays. Composite materials involving the polymer being used with paper or an open-weave fabric, e.g. a glass fibre fabric, are used for gears, bearings and electrical insulation parts.

Amino-formaldehyde materials, generally urea formaldehyde and melamine formaldehyde, give highly cross-linked polymers. Both are used as moulding powders with cellulose and wood flour widely used as fillers. Hard, rigid, high-strength materials are produced with the following table showing some of the properties. Both materials are used for tableware (e.g. cups and saucers), knobs, handles, light fittings and toys. Composites with open-weave fabrics are used as building panels and for electrical equipment.

Epoxide materials are thermosets which are generally used in conjunction with glass (or other) fibres to give hard and strong composites. Polyesters can be produced as either thermosets or thermoplastics. The thermoset form is mainly used with glass (or other) fibres to form hard and strong composites; such composites are used for boat hulls, architectural panels, car bodies, panels in aircraft, and stackable chairs.

Properties of Phenol Formaldehyde Thermostats

Property	Unfilled	Wood Flour Filler
Density (10^3 kg/m^3)	1.25 to 1.30	1.6 to 1.85
Tensile Strength (MPa)	35 to 55	40 to 55
Tensile Modulus (GPa)	5.2 to 7.0	5.5 to 8.0
% Elongation	1.0 to 1.5	0.5 to 1.0
Maximum Service Temperature (°C)	120	150

Properties of Amino-Formaldehyde Thermostats

Property	Urea Formaldehyde Cellulose Filler	Melamine Formaldehyde Cellulose Filler
Density (10^3 kg/m^3)	1.5 to 1.6	1.5 to 1.6
Tensile Strength (MPa)	50 to 80	55 to 85
Tensile Modulus (GPa)	7.0 to 13.5	7.0 to 10.5
% Elongation	0.5 to 1.0	0.5 to 1.0
Maximum Service Temperature (°C)	80	95

Elastomers:

Elastomers are a category of pliable plastic materials that are good at insulating, withstanding deformation, and moulding into different shapes; as a particular kind of polymer, elastomers include natural and artificial rubber. Elastomers are found in a wide variety of applications, from the wheels on a skateboard and the soles of tennis shoes, to the insulation covering speaker cables and telephone lines.

Elastomers are useful and diverse substances that easily form various rubbery shapes. Many industries rely on parts made from elastomers, especially automobiles, sports, electronics, and assembly line factories; this is because these unique polymers offer many unique properties. They are easy to sculpt when they are in their softened, resinous state, yet once they harden, they remain impervious to changes in most changes in temperature as well as stress like stretching or compressing.

Industries working with elastomers have long praised them for being very strong when struck, hard if scratched, resistant to corrosion from various chemicals, and resilient in the face of humidity or water submersion. Since they don't conduct current, they are good electronic insulators; between different branches of wires, they are dense and protective.

Another beneficial property is that they can be compounded or joined with other materials to strengthen certain characteristics. Other kinds of polymers may make them less likely to soften at high temperatures or break down around ozone gas. Elastomers can easily be installed next to various other materials, such as metal, hard plastic or different kinds of rubber with excellent adherence.

The reason that elastomers can deform and return to their previous shape is their cross linked property. Crosslinking means that different chains of polymer molecules have all been linked together, so that the object can uniformly stretch but always returns to its pre-stretching arrangement. One negative aspect of this category of plastics is that they are difficult to recycle, but luckily they last a long time without wearing down.

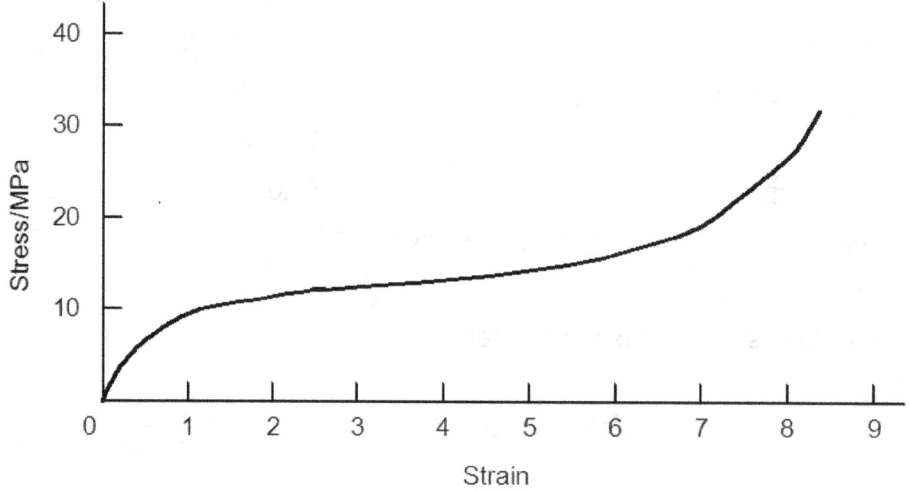

Figure 4.39 – Stress-Strain Graph for an Elastometer

Examples of Elastomers:

The following is a brief outline of elastomers, i.e. rubbers, which are commonly encountered in engineering.

Natural rubber is, in its crude form, just the sap from a particular tree. The addition of sulphur to the rubber produces cross-links, the amount of cross-linkage being determined by the amount of sulphur added. The process of producing cross-links is termed vulcanization. Antioxidants and plasticisers are also added to the rubber.

An example of synthetic rubber is butadiene styrene rubber, commonly called SBR or GR-S or Buna S rubber which is cheaper than natural rubber and is used in the manufacture of tyres, hosepipes, conveyor belts and cable insulation. Another example is butyl rubber, often referred to as isobutylene isoprene or GR-I; this rubber has an important property of extreme impermeability to gases and thus is widely used for the inner linings of tubeless tyres, steam hoses and diaphragms. Nitrile rubbers, known as butadiene acrylonitrile or Buna N, are extremely resistant to organic liquids and are used for such applications as hoses, gaskets, seals, tank linings, rollers and valves. Neoprene, known as polychloroprene, has good resistance to oils and a variety of other chemicals, as well as good weathering characteristics. It is used for oil and petrol hoses, gaskets, seals, diaphragms and chemical tank linings. Polyurethane rubbers have higher tensile strengths, tear and abrasion resistance than other rubbers, are relatively hard and offer good resistance to oxygen and ozone; they are used for oil seals, diaphragms, tyres of forklift trucks and other vehicles where low speeds are involved (not high speeds since they have a low skid resistance), heels and soles of shoes and industrial chute linings.

Properties of Elastomers

Material	Tensile Strength (MPa)	% Elongation	Service Temperature (°C)
Natural Rubber	30	800	-50 to +80
Buna S	24	600	-50 to +80
Butyl Rubber	20	900	-50 to +100
Nitrile Rubber	28	700	-50 to +125
Neoprene	25	1000	-50 to +100
Polyurethane Rubber	36	650	-55 to +125

The Structure of Composites:

A composite material is basically a combination of two or more materials, each of which retains it own distinctive properties. Multiphase metals are composite materials on a micro scale, but generally the term composite is applied to materials that are created by mechanically bonding two or more different materials together. The resulting material has characteristics that are not characteristic of the components in isolation. The concept of composite materials is ancient. An example is adding straw to mud for building stronger mud walls. Most commonly, composite materials have a bulk phase, which is continuous, called the matrix; and a dispersed, non-continuous, phase called the reinforcement. Some other examples of basic composites include concrete (cement mixed with sand and aggregate), reinforced concrete (steel reinforcement bar in concrete), and fiberglass (glass strands in a resin matrix).

In about the mid 1960's, a new group of composite materials, called advanced engineered composite materials (aka advanced composites), began to emerge. Advanced composites utilize a combination of resins and fibres, customarily carbon/graphite, Kevlar, or fiberglass with an epoxy resin. The fibres provide the high stiffness, while the surrounding polymer resin matrix holds the structure together. The fundamental design concept of composites is that the bulk phase accepts the load over a large surface area, and transfers it to the reinforcement material, which can carry a greater load. The significance here lies in that there are numerous matrix materials and as many fibre types, which can be combined in countless ways to produce just the desired properties; these materials were first developed for use in the aerospace industry because for certain application they have a higher stiffness to weight or strength-to-weight ratio than metals. Metal parts can be replaced with lighter weight parts manufactured from advanced composites; generally, carbon-epoxy composites are two thirds the weight of aluminium, and two and a half times as stiff. Composites are resistant to fatigue damage and harsh environments, and are repairable.

Reinforced Concrete *Forming Kevlar* *Submarine Periscope*

Figure 4.40 – Composite Examples

Composites meeting the criteria of having mechanical bonding can also be produced on a micro scale; for example, when tungsten carbide powder is mixed with cobalt powder, and then pressed and sintered together, the tungsten carbide retains its identity. The resulting material has a soft cobalt matrix with tough tungsten carbide particles inside; this material is used to produce carbide drill bits and is called a metal-matrix composite. A metal matrix composite is a type of metal that is reinforced with another material to improve strength, wear or some other characteristics.

Fibres in a Matrix:

The fibres used in a matrix can be continuous long lengths all aligned parallel to an axis of the material, like the steels rods in reinforced concrete, or short fibres randomly orientated in the material. The long length fibres give directionality to the properties, the tensile strength and tensile modulus being much higher along the direction of the fibres than at right angles to them. Randomly orientated short fibres do not lead to this directionality of properties but do not offer such high tensile strengths or tensile modulus values; for example, a glass fibre with reinforced plastic (polyester) might have with long fibres a tensile strength of 800 MPa in the direction of the fibres and only 30 MPa at right angles to them. With short fibres the tensile strength in all directions might be 110 MPa.

Consider a composite rod made up of continuous fibres, all parallel to the rod axis, in a matrix (Figure 4.33). When tensile forces are applied to the composite rod, then each element in the composite has a share of the applied forces.

$$\text{Total Force} = \text{Forces on Fibres} + \text{Force on Matrix}$$

But since stress = force/area then the force on the fibres is equal to the product of the stress s_f on the fibres and their total cross-sectional area A_f. Likewise, the force on the matrix is equal to the product of the stress s_m on the matrix and its cross-sectional area A_m. Therefore:

$$\text{Total Force} = s_f A_f + s_m A_m$$

Dividing both sides of the equation by the total area A of the composite gives:

$$\text{Stress on Composite} = \frac{\text{Total Force}}{\text{Total Area}}$$

$$= s_f \frac{A_f}{A} + s_m \frac{A_m}{A}$$

Thus the stress on the composite is the stress on the fibres multiplied by the fraction of the area that is fibres plus the stress on the matrix multiplied by the fraction of the area that is matrix. Suppose we have glass fibres with a tensile strength of 1500 MPa in a matrix of polyester with a tensile strength of 45 MPa. If the fibres occupy, say, 60% of the cross-sectional area of the composite then the above equation indicates that the tensile strength of the composite, i.e. the stress the composite can withstand when both the fibres and matrix are stressed to their limits, will be:

$$\text{Strength of composite} = 1500 \times 0.6 + 45 \times 0.4 = 918 \text{ MPa}$$

Dividing the stress equation above by this strain gives, since stress/strain is the tensile modulus.

$$\text{Modulus of Composite} = E_f \frac{A_f}{A} + E_m \frac{A_m}{A}$$

The given properties of glass reinforced fibres with a tensile modulus of 76 GPa in a matrix of polyester have a tensile modulus of 3 GPa. If the fibres occupy 60% of the cross-sectional area of the composite then the tensile modulus of the composite is:

$$\text{Modulus of composite} = 76 \times 0.6 + 3 \times 0.4 = 46.8 \text{ GPa}$$

Example
A column of reinforced concrete has steel reinforcing rods running through its entire length and parallel to its axis. If the concrete has a modulus of elasticity of 20 GPa and the steel 210 GPa, what is the modulus of elasticity of the column if the steel rods occupy 10% of the cross-sectional area?

Modulus of composite = 210 x 0.1 + 20 x 0.9 = 39 GPa

Example
Carbon fibres with a tensile modulus of 400 GPa are used to reinforce aluminium with a tensile modulus of 70 GPa. If the fibres are long and parallel to the axis along which the load is applied, what is the tensile modulus of the composite when the fibres occupy 50% of the composite area?

Modulus of composite = 400 x 0.5 + 70 x 0.5 = 235 GPa

Electrical Conductivity:
In terms of their electrical conductivity, materials can be grouped into three categories, namely conductors, semiconductors and insulators. Conductors have electrical conductivities of the order of 10^{-6} S/m, semiconductors about 1 S/m and insulators 10^{-10} S/m. Conductors are metals with insulators being polymers or ceramics. Semiconductors include silicon, germanium, and compounds such as gallium arsenide. Silicon is the most widely used semiconductor.

In discussing electrical conduction in materials, it's useful to visualize an atom as consisting of a nucleus surrounded by its electrons. The electrons are bound to the nucleus by electric forces of attraction. The force of attraction is weaker the further an electron is from the nucleus. The electrons furthest from the nucleus are called the valence electrons since they are the ones involved in the bonding of atoms together to form compounds.

Metals
Metals can be considered to have a structure of atoms with valence electrons which are so loosely attached that they drift off and can move freely between the atoms; therefore, when a potential difference is applied across a metal, there are large numbers of free electrons able to respond and give rise to a current. Electrons can be thought of as pursuing a zigzag path through the metal as they bounce back and forth between atoms. An increase in the temperature of a metal results in a decrease in the conductivity; this is because the temperature rise does not result in the release of any more electrons but causes the atoms to vibrate and scatter the electrons more.

Insulators

Insulators however, have a structure in which all the electrons are tightly bound to atoms; consequently, there is no current when a potential difference is applied because there are no free electrons able to move through the material. To give a current, sufficient energy needs to be supplied to break the strong bonds which exist between electrons and insulator atoms. The bonds are too strong to be easily broken and hence there is no current. A very large temperature increase would be necessary to shake such electrons from the atoms.

Semiconductors

Semiconductors can be regarded as being insulators at a temperature of absolute zero. However, the energy needed to remove an electron from an atom is not very high and at room temperature there can be sufficient energy supplied for some electrons to break free; thus, the application of a potential difference will result in a current. Raising the temperature results in electrons being shaken free which increases the conductivity. When a silicon atom looses an electron it is an electron short and we can consider there to be a hole in its valence electrons. When electrons move they can be thought of as hopping from valence site to a hole in a neighbouring atom, then being released and moving to another hole, etc. Not only do electrons move through the material but so do the holes.

The conductivity of a semiconductor can be very markedly changed by impurities. For this reason the purity of semiconductors must be very carefully controlled. The impurity level of silicon used for the manufacture of semiconductor devices is routinely controlled to less than one atom in a thousand million silicon atoms. Foreign atoms can however be deliberately introduced in controlled amounts into a semiconductor in order to change its electrical properties; this is referred to as doping. Atoms such as phosphorus, arsenic or antimony, when added to silicon, add easily released electrons and so make more electrons available for conduction; such dopants are called donors. Semiconductors with more electrons available for conduction than holes are called an n-type semiconductor. Atoms such as boron, gallium, indium or aluminium add holes into which electrons can move; they are thus referred to as acceptors. Semiconductors with an excess of holes are called a p-type semiconductor.

MEM30007A - Select common engineering materials
Topic 4 - Structure and Properties

Review Problems:
MEM30007-RQ-04

1. Explain the term grain when used in connection with the structure of metals.
2. Explain what is meant by the term alloy.
3. Explain the terms ferrous alloy and non-ferrous alloy.
4. Describe the structure of metals.
5. How does the grain size in a metal affect its properties?
6. How does the shape of grains within a metal affect its properties?
7. Describe the effects on the grain structure and properties of a metal of cold working.
8. Describe the effects on the properties of carbon steels of increasing the percentage of carbon i n the alloy.
9. What types of structure might you expect for a metal which is (a) ductile and (b) brittle?
10. A pure metal is formed into an alloy by larger atoms being forced into the spaces in its crystal structure. What changes might be expected in the properties and why?
11. Describe how the mechanical properties of a cold-worked metal changes as its temperature is raised from room temperature to about $0.6T_m$ where T_m is the melting point on the kelvin scale.
12. How does the temperature at which working is carried out determine the grain size and so the mechanical properties?
13. Why are the mechanical properties of a cold-rolled metal different in the direction of rolling from those at right angles to this direction?
14. How does a cold-rolled product differ from a hot-rolled one?
15. Brasses have recrystallization temperatures of the order of 400°C. Roughly, what temperature should be used for the hot extrusion of brass?
16. A brass, 65% copper and 35% zinc, has a recrystallization temperature of 300°C after being cold worked so that the cross-sectional area has been reduced by 40%. (a) How will further cold working change the structure and properties of the brass? (b) To what temperature should the brass be heated to give stress relief? (c) To what temperature should the brass be heated to anneal it?
17. Refer to Figure 4.27 on page 65. According to this Figure: (a) What is the maximum hardness possible with cold-rolled copper? (b) Copper plate, already cold worked 10%, is further cold worked 20%. By approximately how much will the hardness change? (c) Mild steel is to be rolled to give thin sheets. This involves a 70% reduction in sheet thickness. What treatment would be suitable to give this reduction and a final product which was no harder than 150 HV?
18. Describe the difference between amorphous and crystalline polymer structures and explain how the amount of crystallinity affects the mechanical properties of the polymer.
19. Compare the properties of low and high-density polyethylene and explain the differences in terms of structural differences between the two forms.
20. Why are (a) stabilizers; (b) plasticizers and (c) fillers added to polymers?
21. Describe how the properties of PVC depend on the amount of plasticizer present i n the plastic.

22. Increasing the amount of sulphur in a rubber increases the amount of cross-linking between the molecular chains. How does this change the properties of the rubber?

23. Explain how elastomers can be stretched to several times their length and still be elastic and return to their original length.

24. Calculate the tensile modulus of a composite consisting of 45% by volume of long aligned glass fibres, tensile modulus 76 GPa, in a polyester matrix, tensile modulus 4 GPa. In what direction does your answer give the modulus?

25. In place of the glass fibres referred to in problem 24, carbon fibres are used. What would be the tensile modulus of the composite if the carbon fibres had a tensile modulus of 400 GPa?

26. Long boron fibres, tensile modulus 340 GPa, are used to make a composite with aluminium as the matrix, the aluminium having a tensile modulus of 70 GPa. What would be the tensile modulus of the composite in the direction of the aligned fibres if they constitute 50% of the volume of the composite?

27. How will the properties of composites differ if they are (a) made of long fibres all orientated in the same direction and (b) short fibres with random orientation?

Topic 5 – Processing of Materials:

Required Skills:
- Identify the processing routes for particular materials in order to produce particular microstructures.
- Identify the changes in properties and microstructure associated with processing routes..

Required Knowledge:
- Various types of metals.
- Reading graphs, tables and charts.
- Testing procedures.

Shaping Metals:
Shaping metals concentrates on the structural changes that occur in materials as a result of different manufacturing processes rather than the various shaping methods.

The main methods used to shape metals are:

1. Casting, in which a product is formed by pouring liquid metal into a mould. Sand casting involves using a mould made of sand; die casting uses a metal mould.
2. Manipulative processes, in which a shape is produced by plastic deformation processes. This includes such cold-working methods as rolling, drawing, pressing and impact extrusion. Hot-working processes include rolling, forging and extrusion.
3. Powder techniques, in which a shape which is produced by compacting a powder.
4. Cutting and grinding, in which a shape is produced by metal removal.

The above shaping processes are one way of producing a product; another method is metal joining, of which the main processes are:

1. Adhesives
2. Soldering and brazing
3. Welding
4. Various fastening systems, e.g. rivets, bolts and nuts.

The shaping or assembly method used for a particular product will depend on the metal to be used, its form of supply, and the form of the product. The commercial forms of supply of materials might be as bars, sheet, ingot or pellet.

Casting:
Casting is a manufacturing process where a solid is melted, heated to proper temperature (sometimes treated to modify its chemical composition), and is then poured into a cavity or mould, which contains it in the proper shape during solidification. Consequently, in a single step, simple or complex shapes can be made from any metal that can be melted; the resulting product can have virtually any configuration the designer desires. In addition, the resistance to working stresses can be optimized, directional properties can be controlled, and a pleasing appearance can be produced.

Cast parts range in size from millimetres (such as the individual teeth on a zipper), to in excess of 10 metres and many tonnes (such as the huge propellers and stern frames of ocean liners). Casting has marked advantages in the production of complex shapes, parts having hollow sections or internal cavities, parts that contain irregular curved surfaces (except those made from thin sheet metal), very large parts and parts made from metals that are difficult to machine. Because of these obvious advantages, casting is one of the most important of the manufacturing processes.

Modern technology allows the design of any component to be cast by one or more of the available processes. However, as in all manufacturing techniques, the best results and economy are achieved if the designer understands the various options and tailors the design to use the most appropriate process in the most efficient manner. The various processes differ primarily in the mould material (whether sand, metal, or other material) and the pouring method (gravity, vacuum, low pressure, or high pressure). All of the processes share the requirement that the materials solidify in a manner that would maximize the properties, while simultaneously preventing potential defects, such as shrinkage voids, gas porosity, and trapped inclusions.

The grain structure within the product is determined by the rate of cooling. Figure 5.1 shows the grain structure in a casting. Where the cooling rate is high then small grains are produced, where the rate is low then larger grains are produced; therefore, since the metal in contact with the mould cools faster than that in the centre of the casting, smaller grains are produced at the mould surfaces than in the centre. The small grains produced near the surface are called chill crystals. The cooling rate a little in from the mould walls is less than that at the walls. A consequence of this is that some chill crystals, given time, can develop into long elongated crystals in an inward direction. Such grains are called columnar crystals. The grains produced in the centre of the casting where the cooling rate is the slowest are called equiaxed crystals; these crystals grow in liquid metal which is constantly on the move due to convection currents and as a consequence, the crystals are almost spherical.

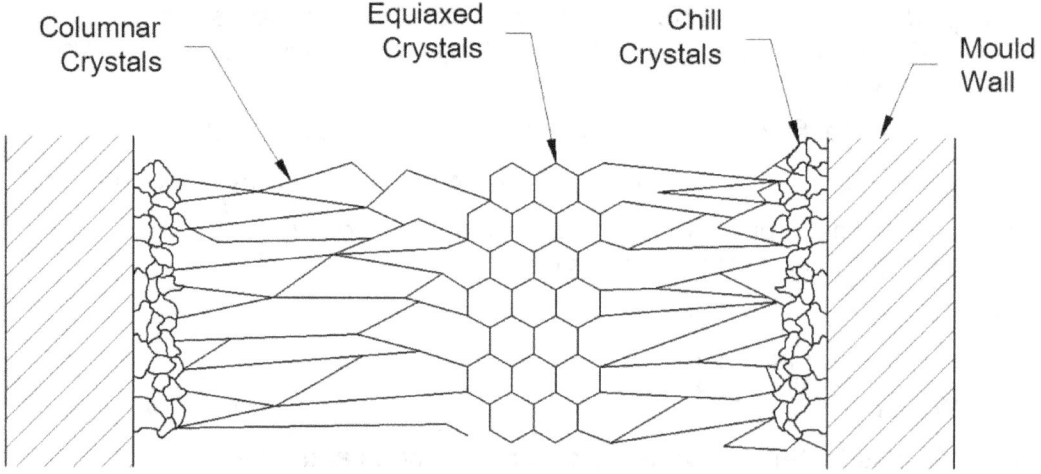

Figure 5.1 – Structure on Solidification of an Ingot

Castings in which the mould is made of sand tend to have a slow rate of cooling as sand has a low thermal conductivity; hence sand castings tend to have large columnar grains. Since large grains mean a low strength and hardness then sand castings have relatively low strength and hardness. Die castings involving metal moulds have a much faster rate of cooling and so give castings for which there has not been enough time for long columnar crystals to develop and therefore have a bigger zone of equiaxed crystals. As these are relatively small crystals then the casting has better mechanical properties than the corresponding sand casting.

Example

What type of grain structure might be expected when liquid metal is poured into a narrow metal mould?

The rate of cooling will be high because it is a metal mould; also, because the mould is thin, all parts of the casting will cool quickly therefore it is likely that chill crystals will be formed throughout since there is not enough time for columnar crystals to develop.

Manipulative processes:

Manipulative methods involve the shaping of a material by plastic deformation processes. The products given by such methods are said to be wrought. Where the deformation is carried out below the recrystallization temperature, the process is said to involve cold-working, when in excess of that temperature it is the hot-working process. The main cold-working processes are cold rolling, drawing, pressing and impact extrusion; the main hot-working processes are rolling, forging and extrusion.

Cold-worked metals generally have deformed grains and consequently are harder and more brittle, being said to be work hardened; they also have a directionality of properties since the grains are deformed in the direction of the manipulation, consequently, with cold rolling, the sheet in the direction of the rolling has different properties at right angles to that direction. Cold rolling may have to take place in a number of stages with annealing between the stages because the metal becomes too work hardened for the entire reduction in thickness to occur with one operation. The annealing results in the grains becoming large and undistorted. Drawing involves the pulling of metal through a die; Figure 5.2 illustrating this process for wire drawing. Deep drawing in Figure 5.3 involves sheet metal being pushed through an aperture by a punch. The more ductile metals such as aluminium, brass and mild steel can be used to shape products by this method. Both the drawing or wires and deep drawing result in the material work hardening and assuming a directionality of properties. Hot-worked metals have larger grains and consequently are softer and more ductile. Hot rolling is generally at a temperature of about $0.6T_m$, where T_m is the melting point of the metal on the Kelvin scale. At this temperature, work hardening does not occur though the surfaces of the material become oxidized. Forging involves squeezing by pressing or hammering, a ductile metal between a pair of dies so that it ends up assuming the internal shape of the dies. Figure 5.4 illustrates this process in relation to what is termed closed-die forging. Generally, for the material to be ductile enough during the operation the forging has to take place at a temperature in excess of the recrystallization temperature. The flow of the material during the squeezing operation does give some directionality to the properties of the material with grains and any non-metallic inclusions in the metal tending to end up aligned along the directions of flow.

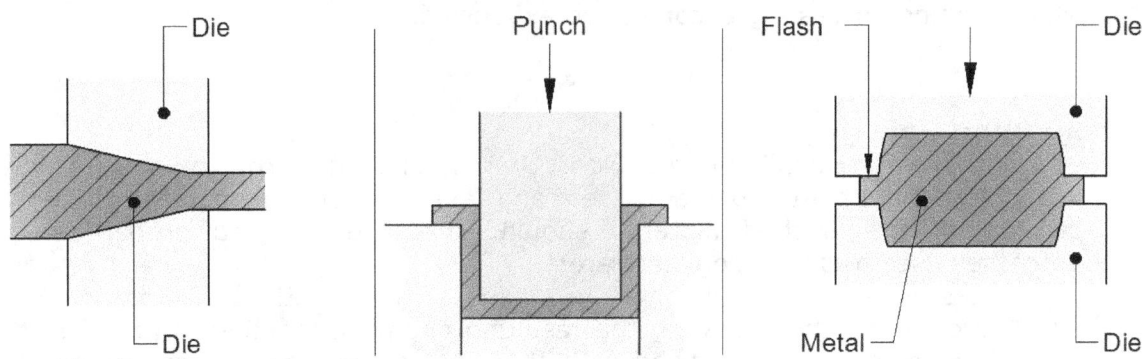

Figure 5.2 – Drawing a Wire   *Figure 5.3 – Deep Drawing*   *Figure 5.4 – Closed-Die Forging*

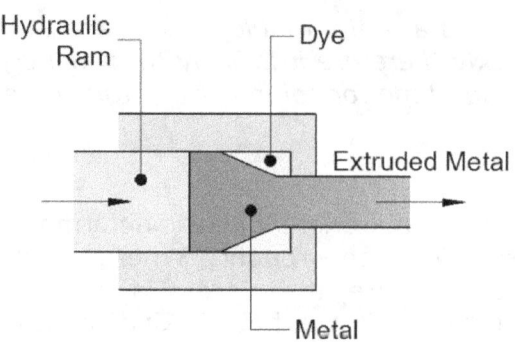

Figure 5.5 - Extrusion *Figure 5.6 – Sintered Bronze Bush*

Extrusion is rather similar to the squeezing of toothpaste out of a tube. The shape of the extruded toothpaste is determined by the nozzle through which it is ejected. Figure 5.5shows the basic principles. Cold extrusion is when the extrusion takes place below the recrystallization temperature, hot extrusion at a higher temperature. Typically, hot extrusion takes place at a temperature of the order of 0.65 to $0.9T_m$. Cold extrusion gives a work-hardened product, hot extrusion a soft, ductile, one. Hot extrusion is generally required when the cross-section of the material is reduced by a factor of 50 or more. The flow of the material that occurs during extrusion ends up in giving some directionality of grains and hence properties.

Example
What types of carbon steels can be used with the deep drawing process?

For deep drawing high ductility is required. As a consequence, only carbon steels with low amounts of carbon can be used as only they have sufficient ductility (refer to Figure 4.9 – Properties of Carbon Steels and the associated text); accordingly, mild steel can be used but not a high-carbon steel.

Powder Techniques:
Shaped metal components can be produced from a metal powder. The process, called sintering, involves compacting the powder in a die and then heating it to a sufficiently high temperature for the particles to knit together. The compacting forces particles into contact with each other and then the heating enables atoms in particles to diffuse across the points of contact and form necks which hold the particles together when the material cools. Sintering is a useful method for the production of components from brittle materials where manipulative processes are difficult and high melting point materials for which melting for casting becomes too expensive.

Shaping Polymers:
Polymers may be supplied in a powder, granule or sheet form, the supplier having mixed the polymer with suitable additives and even other polymers in order that, after processing, the finished material should have the required properties. The main processes used to shape polymers are:
1. Casting – This can involve the mixing of the constituent parts of the plastic in a mould and then allowing the resulting chemical reaction to produce the polymer. The method can be used with thermosets and thermoplastics; another method involves the melting of the powdered polymer in a heated mould.
2. Moulding – With thermoplastics the polymer might be melted and forced into a mould, the process being called injection moulding. With thermosets, the powdered polymer may be compressed between the two parts of the mould and then heated under pressure; the process is known as compression moulding. With transfer moulding the powdered thermoset is heated in a chamber before being transferred by a plunger into the mould.
3. Forming – A polymer sheet, or thermoplastic, is heated and pressed into or around a mould.

4. Extrusion – Where a thermoplastic polymer is forced through a die.

In addition, products may be formed by polymer joining. The main processes are:
1. Adhesives
2. Welding
3. Various forms of fastening systems, e.g. riveting, press and snap fits, screws.

Flowing Processes:
Many of the polymer processes used to produce products can be considered to involve the flowing of liquid polymer and its subsequent cooling in the required shape. Figure 5.7 shows the basic principle of one form of injection moulding, a very widely used process where the polymer is melted and forced into a mould. The method is capable of giving high production rates because the rate of cooling of the polymer in the mould is fairly fast.

Consider a linear chain thermoplastic material, which is capable of crystallizing. While the polymer is in the liquid state the molecules are able to move about. For chains of the polymer to form, there has to be sufficient time for the molecules being incorporated into the chain to arrive at the correct positions in the chain; consequently starting with a small length of chain, the chain steadily increases in length as more and more molecules arrive at its ends and join on. We can consider that some of these chains, while growing, fold to give a crystal; these crystals can then grow as more molecules attach themselves to the ends of the chain. All this, however, takes time; the size of the polymer crystals, whether there is crystallinity at all, depends very much on the time during which the polymer is liquid. Hence it depends on the rate of cooling. Large crystals can be produced by slow cooling, small crystals by fast cooling; if there is very fast cooling then no crystallinity may occur and the polymer is completely amorphous.

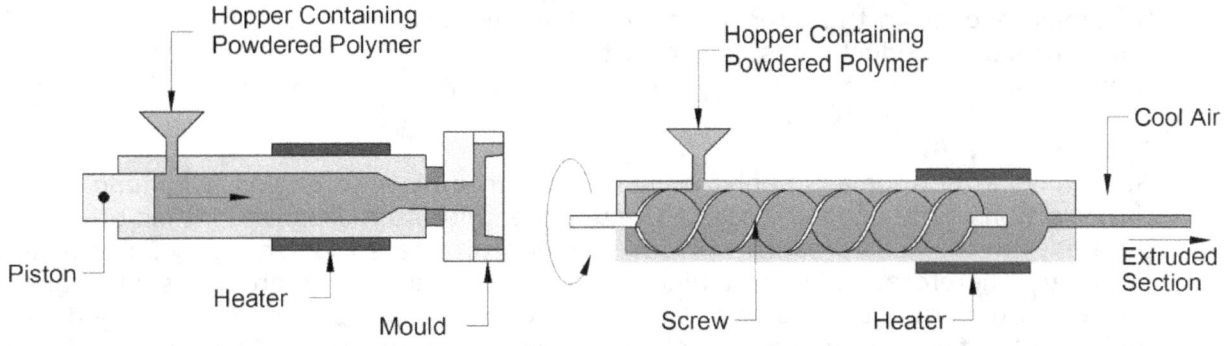

Figure 5.7 – Injection Moulding   *Figure 5.8 – Extrusion*

Because injection moulding has a fast cooling rate, necessary for high production rates, it can result in a polymer, which could be crystalline at slow rates of cooling, being completely amorphous which has consequences for the properties of the material. Amorphous polymers can be stretched more than crystalline polymers; for some polymers, the rate of cooling is slow enough for the inner part of the moulded shape to crystallize while the outer layers, which cool faster than the inner part, are amorphous. As a consequence, because crystalline polymer chains are more tightly packed than randomly arranged ones, the inner part has a higher density than the outer layers.

With injection moulding, liquid polymer is forced into the mould. The direction of flow of the polymer usually varies with position throughout the mould; since the molecules tend to align with the direction of the flow, the plastic product can have mechanical properties which vary with direction.

Extrusion involves the forcing of liquid polymer, a thermoplastic, through a die. The process is comparable with the squeezing of toothpaste out of its tube. Figure 5.8 shows the basic form of the extrusion process. The polymer is fed into a screw mechanism which takes it through the heated zone and forces it out through the die where it cools rapidly. As a consequence, the cooling might be too fast for crystals to develop, so an amorphous polymer product results. In some cases the inner core of the extruded polymer might cool slowly enough for some crystallinity, while just the outer layers remain amorphous. The direction of the polymer chains is also likely to follow that of flow of the polymer and so give a product with different properties along the length of the extrusion than at right angles.

Example

One way of producing sheet plastic is to extrude the polymer through a slit. What types of properties might be expected of the sheet?

The extrusion process will tend to align the polymer molecules with the direction of flow. Thus the sheet might be expected to have different properties in directions along the length of the sheet and at right angles.

Manipulative Processes:

Forming processes are used to form articles from sheet thermoplastic polymer with the heated sheet being pressed into or around a mould. The term thermoforming is often used. The sheet may be pressed against the mould by the application of pressure on the sheet; this method is called thermoforming. Alternatively, it can be by the production of a drop in pressure between the sheet and the mould, as illustrated in Figure 5.8; the method is called vacuum forming. Forming processes involve a polymer being stretched; as a consequence, the molecular chains are forced into becoming aligned in the direction of the stretching. The formed product has a directionality of properties.

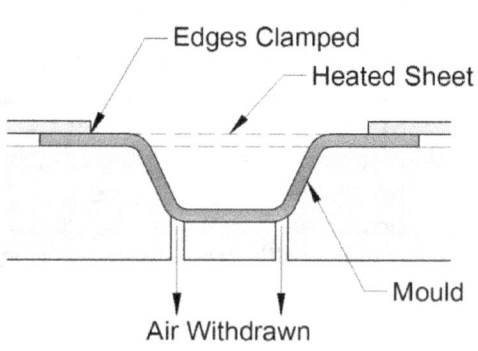

Figure 5.9 – Vacuum Moulding

Drawing of Polymers:

Stretching a thermoplastic polymer can result in molecular chains becoming lined up in the direction of the applied forces. The polymer with its molecules all thus orientated has different mechanical properties from the non-orientated polymer, being considerably stronger; therefore, cold stretching such a polymer can improve its strength. The process is generally termed cold drawing, however, if the polymer is stretched while hot and then the forces removed it is still possible for the molecules to become disorientated during the cooling and no such alignment of molecules and improvement in properties occurs. Hot stretching has virtually no effect on the properties. The plastic bottles used for soft drinks, e.g. Coca-Cola, are made from polyethylene terephthalate (PET). The method uses injection moulding to produce an initial bottle-shape. If the PET material is cooled quickly after the moulding then the material is amorphous, if cooled slowly crystalline. The amorphous material is transparent, permeable to the 'fizz' and not very strong. The crystalline material is opaque, impermeable and stronger; what is wanted, however, is the combination of transparency with impermeability and strength. The procedure that is adopted is to first obtain the amorphous form. The bottle is then heated, but not so high that plastic flow can occur, and stretched in length and generally expanded to the required shape by air pressure blowing the bottle out to its full size in another mould. The process results in some molecular alignment occurring with a consequent improvement in properties while still retaining the transparency.

Example

Nylon thread is often cold drawn after its production. What effect is this likely to have on the structure and properties of the nylon?

The cold drawing improves the alignment of the polymer molecules and so results in improved strength.

Heat Treatment of Metals:

Heat treatment can be defined as the controlled heating and cooling of metals i n the solid state for the purpose of altering their properties. A heat treatment cycle consists normally of three parts:
1. Heating the metal to the required temperature for the changes i n structure with in the material to occur.
2. Holding at that temperature for long enough for the entire material to reach the required temperature and the structural changes to occur throughout the entire material.
3. Cooling, with the rate of cooling being controlled since it affects the structure and hence properties of the material.

Annealing is the heat treatment used to make a metal softer and more ductile; it involves heating the metal to above the recrystallization temperature and then slow cooling. The result is a regrowth of grains to give a large grain structure.

In the case of carbon steels, this change in grain size is also accompanied by changes in the form of the constituents present in the alloy. Heating steel to above the recrystallization temperature changes the crystal structure from ferrite to austenite. The austenite can contain more carbon atoms than the ferrite; there is time with annealing, because the cooling is slow, for the excess carbon atoms to move out of the crystal structure to form a compound called cementite. If the heated metal is cooled very rapidly, for example by being dropped into cold water, there is not time for the austenite to lose its excess carbon atoms and they become trapped in the structure. The result is a new structure called martensite where the carbon trapped in this structure considerably distorts the structure. As a consequence, martensite is very hard and brittle and the steel becomes harder and more brittle. The hardness and strength increases quite significantly with an increase in carbon content, this being because more carbon is trapped in the structure and so there is more distortion and is called quenching.

A problem with the severe cooling occurring in quenching is that cracks can occur which are a result of the distortion produced by the structural changes and also differential expansion as a result of different parts of a product cooling at different rates. Figure 5.10 shows how the hardness of carbon steels after quenching compares with that of the annealed steels.

In the quenched state, steels have such a low ductility as to be very difficult to use. The process known as tempering can, however, be used to improve the ductility without losing all the hardness gained by the quenching.

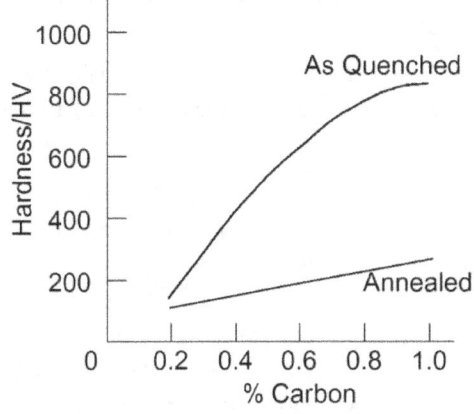

Figure 5.10 – Hardness of Carbon Steels

Tempering involves heating the steel to a temperature at which some of the carbon trapped in the martensite structure can diffuse out and form cementite, so reducing the distortion of the structure. The amount of carbon that diffuses out depends on the temperature used for the tempering. The mechanical properties depend on the tempering temperature. **Error! Reference source not found.** shows this for alloy steel (manganese-nickel-chromium- molybdenum steel).

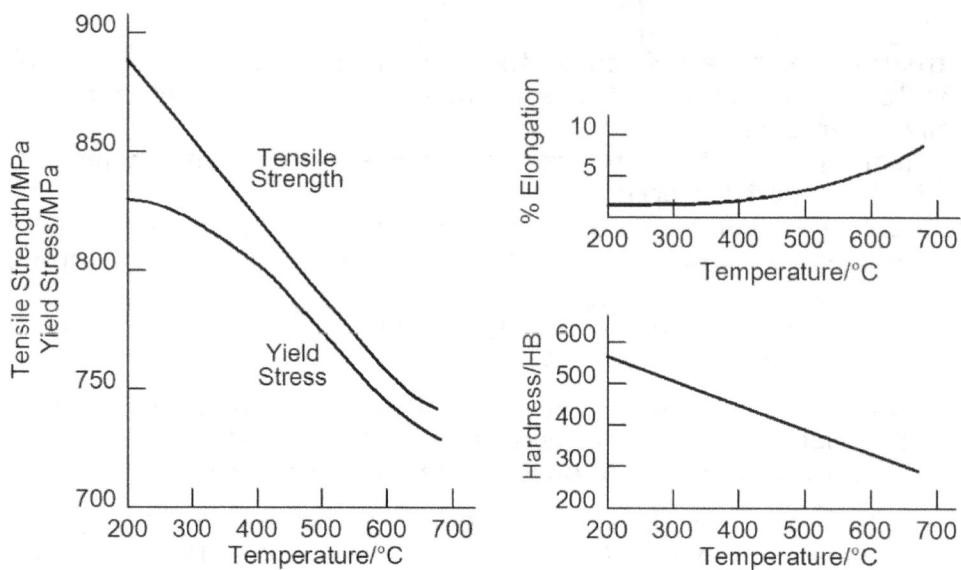

Figure 5.11 – The Effect of Tempering

A wide range of alloys used in engineering depend on a treatment called precipitation hardening for improvements in their hardness and strength; this type of treatment is widely used with aluminium alloys and nickel alloys. The process involves heating the alloy to above the recrystallization temperature, then quenching resulting in a distorted crystal structure; however, with time, atoms diffuse out of the structure of this type of alloy to give a fine precipitate which precipitate lodges at grain boundaries and in dislocations and, as a consequence, makes slip more difficult. The result is an increase in hardness and strength, for example, the aluminium-copper alloy (3003) might have a tensile strength of 185 MPa and hardness 45 HB in the annealed state and after precipitation hardening, 425 MPa and 105 HB.

Example

The following are mechanical properties of a nickel alloy. Identify the internal structure which is responsible for the properties:

	Strength (MPa)	Yield Strength (MPa)	Elongation (%)
Cold Worked	535	380	12
Cold Worked and Annealed	380	105	36

The cold working has caused grains to be become distorted and include a greater number of dislocations. As a consequence, the material is strong and relatively brittle. Annealing the material has removed the distortion, increasing the grain size and reducing the number of dislocations. As a result, the material is weaker and more ductile.

Example

The following are the mechanical properties of a carbon steel. Identify the internal structure which is responsible for these properties:

	Strength (MPa)	Yield Strength (MPa)	Elongation (%)
As Rolled	815	240	17
Annealed	625	180	23
Quenched and Tempered at 200°C	1100	320	13
Quenched and Tempered at 650°C	800	230	13

The as-rolled material will have some work hardening and therefore distorted grains with increased numbers of dislocations. Annealing the material results in recrystallization and larger grains being produced with a drop in the number of dislocations. Therefore the material is weaker, softer and more ductile. Quenching results in the formation of martensite in which carbon atoms are trapped within a distorted structure. As a consequence, the material is much harder and more brittle. Tempering allows some of the carbon atoms to diffuse out and reduce the distortion, the higher the tempering temperature, the greater the reduction. Hence tempering restores some of the ductility but reduces the strength.

Surface Hardening:

Surface hardening is the treatment of steel by heat or mechanical means to increase the hardness of the outer surface while the core remains relatively soft. The combination of a hard surface and a soft interior is greatly valued in modern engineering because it can withstand very high stress and fatigue, a property that is required in such items as gears and anti-friction bearings. Surface-hardened steel is also valued for its low cost and superior flexibility in manufacturing.

Many situations in industry require the need for the surface of a piece of steel to be hard, e.g. to make it wear resistant, without the entire component being made hard and often too brittle. Several methods are available for surface hardening:

Flame Hardening:
For carbon steels surface hardening can be achieved by just heating the surface layers to above the recrystallization temperature and then quenching to give a martensitic structure for these surface layers. The selective heating of the surface layers can be carried out with an oxyacetylene flame.

Induction Heating:
Induction heating is a process which is used to bond, harden or soften metals or other conductive materials. For many modern manufacturing processes, induction heating offers an attractive combination of speed, consistency and control.

The basic principles of induction heating have been understood and applied to manufacturing since the 1920s. During World War II, the technology developed rapidly to meet urgent wartime requirements for a fast, reliable process to harden metal engine parts. More recently, the focus on lean manufacturing techniques and emphasis on improved quality control have led to a rediscovery of induction technology, along with the development of precisely controlled, all solid state induction power supplies.

The steel component is placed in a coil carrying a high-frequency current and allowing the induced currents in the surface layers to do the heating.

Case Hardening:
Case hardening is a simple method of hardening steel and is a less complex process than hardening and tempering. The technique is used for steels with low carbon content. Carbon is added to the outer surface of the steel, to a depth of approximately 0.03mm. One advantage of this method of hardening steel is that the inner core is left untouched and so still processes properties such as flexibility and is still relatively soft.

The process is achieved by heating the steel while it is packed in charcoal and barium carbonate or in a furnace in an atmosphere of a carbon-rich gas or, alternatively, in a bath of liquid sodium cyanide. The methods may result in steel having an inner core containing 0.2% carbon and surface layers with 0.9% carbon

Nitriding Carbonitriding:
During the carbonitriding process, atoms of carbon and nitrogen diffuse interstitially into the metal, creating barriers to slip, increasing the hardness and modulus near the surface. Carbonitriding is often applied to inexpensive, easily machined low carbon steel to impart the surface properties of more expensive and difficult to work grades of steel. Surface hardness of carbonitrided parts ranges from 55 to 62 HRC.

Certain pre-industrial case hardening processes include not only carbon-rich materials such as charcoal, but nitrogen-rich materials such as urea, which implies that traditional surface hardening techniques were a form of carbonitriding.

Carbonitriding is similar to gas carburization with the addition of ammonia to the carburizing atmosphere, which provides a source of nitrogen. Nitrogen is adsorbed at the surface and diffuses into the workpiece along with carbon. Carbonitriding (around 850°C/1550°F) is carried out at temperatures substantially higher than plain nitriding (around 530°C/990°F) but slightly lower than those used for carburizing (around 950°C/1700°F) and for shorter times. Carbonitriding tends to be more economical than carburizing, and also reduces distortion during quenching. The lower temperature allows oil quenching, or even gas quenching with a protective atmosphere.

Integrated Circuit Fabrication:

A typical integrated circuit is a chip about 5 mm across by 1 mm thick and has been cut from a wafer of semiconductor material which contains hundreds of such chips. An integrated circuit contains a complete circuit made up from such components as transistors, diodes, resistors and capacitors. The components are fabricated by selective doping of the surface layers of the chip, the components being formed in the top 10 to 20 μm. The substrate is a p-type semiconductor with the components in the surface layers formed by the selective introduction of n and p-type dopants. The surface of die chip is covered with silicon dioxide, which is an electrical insulator, with connections to the underlying components being made by aluminium which is deposited over the silicon dioxide and through holes (termed windows) in this layer. The layer of silicon dioxide can be produced by a chemical reaction between the wafer surface and either oxygen or steam, or by a deposition process. Etching is used to remove unwanted regions of material from a wafer, e.g. cutting windows in the silicon dioxide layer through which doping of the underlying layers can occur. The etching can be done by:

1. Wet etching, in which wafers are immersed in an acid bath

2. Plasma etching, with the surface being exposed to reactive gas atoms, generated by the breakdown of a gas, as a result of being heated by radio-frequency electromagnetic energy

3. Ion milling, in which a beam of high energy ions are used as a bombarding beam to dislodge atoms.

Dopants can be introduced, through windows, into a wafer by solid diffusion or ion implantation. With solid diffusion, the dopant atoms are deposited on the surface and the wafer is then heated. The atoms gradually diffuse into the surface layers. With ion implantation, the dopant is fired at the wafer as a beam of ions.

Topic 5 - Processing of Materials

Review Problems:

MEM30007-RQ-05

1. Explain why copper wires that are used for electrical conductors are usually finished by a cold-drawing process and then heated to about 700°C.

2. What microstructure, and hence what properties, would you expect in cold-drawn wire if there is no further treatment of it?

3. Which type of casting, sand or die casting will produce a product with the smallest grains?

4. With a thermoplastic polymer, how does the rate of cooling from the liquid state affect the degree of crystallinity?

5. What would you expect to be the internal structure of an extruded thermoplastic?

6. The thin plastic containers used to hold biscuits or chocolates in boxes are produced by thermoforming a thermoplastic. What is likely to be the resulting molecular structure in the various parts of the container?

7. Polythene bags are generally made by extruding the polyethylene through a die to give an extruded cylinder. The cylinder, while hot, is then inflated by air pressure. What type of structure might be expected within the bag material? You might care to try an experiment of cutting strips in directions along the length of the bag and at right angles and pulling them between your hands. The results can then be compared with the structure predicted by your answer.

8. Polypropylene twine consists of fibres of polypropylene which have been cold drawn. What is the effect of this process?

9. State what structural changes take place and the consequential changes in properties: in (a) annealing; (b) quenching; (c) tempering; (d) precipitation hardening; (e) flame hardening and (f) case hardening.

10. Carbon steel is found to have the following properties. Explain how they arise in terms of the structure of the steel.

	Strength (MPa)	Yield Strength (MPa)	Elongation (%)
As Rolled	550	180	32
Annealed	465	125	32
Quenched and Tempered at 200°C	850	495	17
Quenched and Tempered at 650°C	585	210	31

11. An aluminium-manganese alloy is found to have the following properties. Explain how they arise in terms of the structure of the alloy.

	Strength (MPa)	Yield Strength (MPa)	Elongation (%)
Cold Worked	110	28	30
Cold Worked and Annealed	200	55	4

12. An aluminium-magnesium-silicon alloy is found to have the following properties. Explain how they arise in terms of the structure of the alloy.

	Strength (MPa)	Yield Strength (MPa)	Elongation (%)
Cold Worked	125	30	25
Cold Worked and Annealed	310	95	12

13. The striking part of the head of a hammer is required to be very hard, but the main body of the hammer head must be softer and tougher. How can these properties be achieved in a single piece of steel?

14. What properties are required of a hacksaw blade and how might they be obtained?

Topic 6 – Selection of Materials:

Required Skills:
- Identify the materials properties required for a particular specification/application.
- Recognize materials with the required properties.
- Suggest processing routes.
- Discuss the behaviour of materials in service.

Required Knowledge:
- Various types of metals.
- Reading graphs, tables and charts.
- Testing procedures.

Requirements:
Before selecting any material for a component the question to be asked is "What functions does a product have to perform?" The question is an important because it requires an answer before either the materials or the forming processes for the materials are considered; from this stems a sequence of further questions. The following case study serves to illustrate how the sequence might develop.

Consider the problem of making a domestic kitchen pan. Its functions may be deemed to be to hold liquid and allow it to be heated to temperatures of the order of 100°C. By considering the function we can arrive at the basic design requirements, therefore, a consequence of these functions for the pan are:
- Particular shape for the container.
- Container must not deform when heated to these temperatures.
- Good conductor of heat.
- Leakproof.
- Not ignite when in contact with a flame or hot electrical element.
- Have a non-stick surface.
- An attractive finish.

From these requirements we can now define the required properties of the materials. Thus for the pan, the requirement that the material be a good conductor of heat would seem to reduce the consideration to metals, particularly when taken together with the requirement that the material can be put in contact with a flame and contain hot liquids which would effectively rule out polymers. But what properties are required of the metal? The shape of the pan would suggest that a deep-drawing process be used; as this is a cold-working process then there will be a good surface finish. For deep drawing the initial material must be reasonably ductile and available in sheet form hence we might consider an aluminium alloy. Referring to the tables, a possibility would seem to be an aluminium-magnesium alloy, 5005. Another possibility would be a stainless steel, stainless because rusty pans would not be very desirable. Again, referring to the appropriate tables, a possibility would seem to be 302S31; the deciding factor is likely to be cost, though there may be some prestige value attached to having stainless steel plans as opposed to aluminium, which would allow a higher price to be charged. For the same volume of material, the stainless steel will probably cost about three times the aluminium alloy.

The above represents one line of argument regarding the design of pans. It is instructive to examine a range of pans and consider the materials used and what reasons might be advanced for them being chosen. Why, for example, are some pans made of glass, of a ceramic, of a steel coated on the outside with an enamel on the inside with a non-stick polymer polytetrafluoroethylene (PTFE)?.

The above is only the consideration of the container part of the pan; there is still the handle to study. The function required is that it can be used to lift the pan and contents, even when they are hot. The properties required are:
- Poor thermal conductivity.
- Able to withstand the temperatures of the hot pan.
- Stiffness and adequate strength.

The handle can be considered to be essentially a cantilever with a load, the pan and contents, at its free end. Before going too far in considering the design and materials for the handle; taking this into consideration, then the need for the handle to have low thermal conductivity indicates that metal would not be a good choice. The possibility is thus a polymer. It needs, however, to be able to withstand a temperature of the order of 100°C at the pan end and have a reasonably high modulus of elasticity and reasonable strength; these requirements suggest that a thermoset is more likely to be feasible than a thermoplastic. A possibility is Phenolic resin, micarta or PP. When filled with wood flour as an example it has a high enough maximum service temperature of about 150°C, a tensile modulus of 5.0 to 8.0 GPa (high for a polymer), and a tensile strength of 40 to 55 MPa. Because it is a thermoset then the processing method could be casting.

In the above considerations of the pan and the handle the item that has so far not been discussed is the life of the items. The purchaser would expect the pan to last without problems for a reasonable period of time which is likely to be years. The handle should not break during this time or discolour or deteriorate when used and washed a large number of times. The pan should not wear thin or change its mechanical properties with frequent heating, exposure to hot liquids and washing-up liquids.

Stages in the selection process

As the above examples indicate, there are a number of stages involved in arriving at possible materials and processing requirements for a product. These can be summarised in very simple terms, as follows:
- Define the functions required of the product.
- Consider a tentative design, taking into account any codes of practice, national or international standards.
- Define the properties required of the materials.
- Identify possible materials, taking into account availability in the required forms.
- Identify possible processes which would enable the design to be realized.
- Consider the possible materials and possible processes and arrive at a proposal for both. If not feasible, consider the design again and go back through the cycle.
- Consider how the product will behave during its service life

Costs:

The total cost to the consumer of a manufactured article in service, i.e. the so-called *total life cost,* is made up of a number of items; these are:
- The purchase price – includes the costs of production, the fixed costs arising from factory overheads, administration, etc. and the manufacturer's profit. The costs of production include the cost of the materials and the cost of manufacture.
- The cost of ownership – includes such costs as those associated with maintenance, repair and replacement.

Failure in Service:

The failure in service of structures, equipment and componentry attributes to large loss of time and costly expense to industry. Catastrophic failures have led to too many fatal accidents and long-term alterations to public and private infrastructure. Some examples of such failures are the collapse of the 12-Storey Building Collapse (2013 - Egypt) and West Gate Bridge (1970 - Australia), Chernobyl Reactor Disaster (1986 – Russia), Exxon Valdez Oil Spill (1989 – Alaska), Italian cruise ship 'Costa Concordia' (2012 – Italy); all the examples resulted in loss of life.

Not all failures cause loss of life but can have a great impact of national services such as the Blackout of North-Eastern USA and Canada in 2003 or a major part of the Sydney railway system grinding to a halt in 2013 due to a fire warning light in a signal box.

All failures must be eliminated entirely from the workplace and infrastructures to reduce loss of life and escalating expenses. Failures in service can arise from:
- Errors in the original design, e.g. the wrong material having been chosen.
- The material used is in some way defective, e.g. the specification of the material to be obtained from the suppliers was not tight enough or was below specification and not detected by inspection.
- Defects are introduced during the manufacturing process, e.g. heat treatment gives cracks as a result of quenching or perhaps incorrect assembly leading to misalignment and high stresses.
- Deterioration in service, e.g. the product is exposed to unexpected corrosive environments or perhaps a temperature which results in changes in the microstructure of the material or perhaps poor maintenance which leads to nicks and gouges which act as stress raisers and a greater chance of failure due to fatigue.

The causes of Failure:

The causes of failure are vast and nearly infinite however the following are the more common causes of failure.

Human Error:
Humans and not perfect and therefore **"accidents do not happen, but are caused"** by complacency, inattention, rushing to finish and sheer carelessness. All work should be taken seriously and checked by other workers and supervisors. Human error could involve under tightening of fastenings, failure to remove tape from a venturi inlet, untidy workplace or not carefully undertaking the calculation of a formula.

Over Stressed/Under Designed:
The stress level is just high and the material yields and breaks. The material may show a ductile or a brittle form of fracture; with a ductile fracture a significant yielding will occur before the actual fracture while brittle fracture virtually have none.

Fatigue:
The material is subject to an alternating stress which results in fatigue failure. Bending a stiff piece of metal or plastic to repeat flexing back and forth and subjecting it to an alternating stress going from tension to compression to tension to compression and so on will ultimately cause fatigue failure and is generally an easier way of causing the material to fail than applying a direct pull. The chance of fatigue failure occurring is increased the greater the amplitude of the alternating stresses. Stress concentrations produced by holes, surface defects and scratches, sharp comers, sudden changes in section, etc. can all help to raise the amplitude of the stresses at a particular point in the material and reduce its fatigue resistance. An example of fatigue stress is the de Havilland Comet crash in 1954.

Creep:
The material is subject to a load which initially does not cause failure but the material gradually extends over a period of time until it fails and is known as creep. For most

metals creep is negligible at room temperature but can become pronounced at high temperatures; for plastics, creep is often quite significant at ordinary temperatures and even more pronounced at higher ones. The creep of a metal, or a polymer, is determined by its composition and the temperature; for example, aluminium alloys will creep and fail at quite low stresses when the temperature rises above 200°C, while titanium alloys can be used at much higher temperatures before such a failure occurs.

Shock Loading:
A suddenly applied load causes failure, i.e. the energy at impact is greater than the impact strength of the material. Brittle materials have lower impact strengths than ductile ones.

Environmental Changes:
The temperature changes and causes the properties to change in such a way that failure results, e.g. the temperature of a steel may drop to a level at which the material turns from being ductile to brittle and then easily fails as a result of perhaps an impact load.

A temperature gradient is produced and causes part of the product to expand more than another part, the resulting stresses resulting in failure, e.g. if hot water is poured into a cold glass then the inside of the glass tries to expand while the outside does not (glass has a low thermal conductivity), the result being that the glass cracks.

Expansion:
The product is made of materials with differing coefficients of expansion. Thus when the temperature rises the different parts expand by different amounts, with the result that internal stresses are set up which can result in distortion and possible failure.

Thermal Cycling:
Thermal cycling in which the temperature of the product repeatedly fluctuates will result in cycles of thermal expansion and contraction. If the material is constrained in some way then internal stresses will be set up and, as a consequence, the thermal cycling will result in alternating stresses being applied to the material. Fatigue failure can result and is referred to as thermal fatigue.

Degradation:
Degradation because of the environment in which the material is situated, e.g. corrosion of iron leading to a reduction in the cross-section of a product and hence the resulting increase in stress leading to failure. Corrosion prevention, e.g. painting iron, is a major way of avoiding this type of problem. Plastics may become brittle as a result of exposure to the ultraviolet radiation in sunlight; this effect can be reduced by incorporating stabilizers with the polymer.

Examination of Failures:
A ductile material is characterized by having a significant plastic region to its stress-strain graph; when a ductile material has a gradually increasing tensile stress applied, then when yielding starts, the cross-sectional area of the material becomes reduced with necking being said to occur. Eventually after a considerable reduction in cross-sectional area the material fails. The resulting fracture surfaces show a cone and cup formation which occurs because, under the action of the increasing stress, small internal cracks form which gradually grow in size until there is an internal, almost horizontal, crack. The final fracture occurs when the material shears at an angle of 45° to the axis of the applied stress; the type of failure is referred to as a *ductile fracture*. Materials can also fail in a ductile manner in compression, such fractures resulting in a characteristic bulge and series of axial cracks around the edge of the material.

In Figure 6.1 a sample test piece has an initial strain applied pulling each ends apart; the strain increases causing necking to occur towards the centre of the sample. An internal crack forms as the elastic limit is reached and after the maximum stress is exceed, the test piece fails and parts along the crack. A close inspection of the fracture shows one part has a cone formed with an angle of approximately 45° while a hollow cup is moulded in the other part.

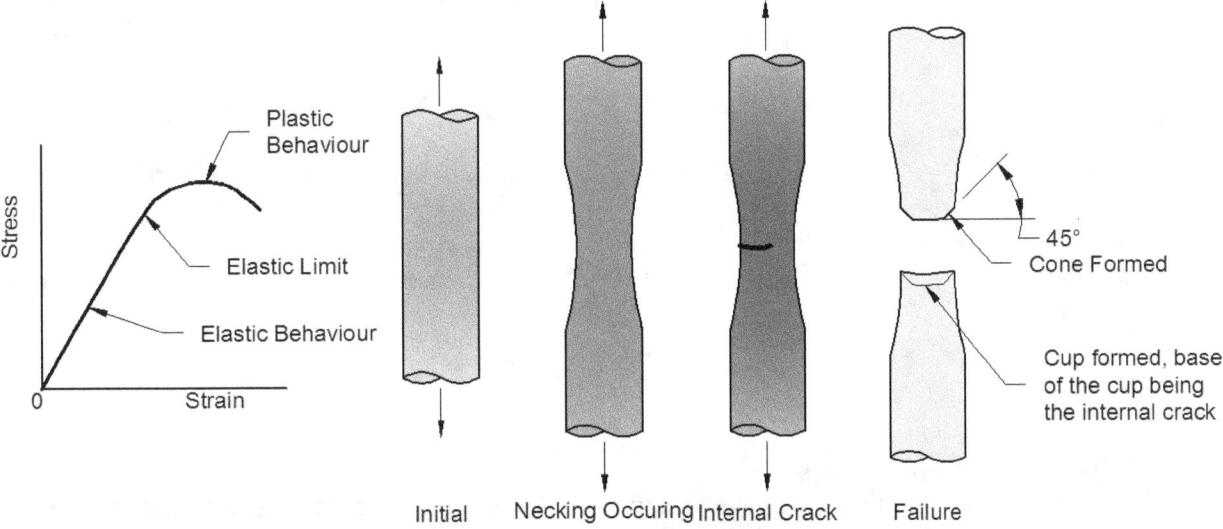

Figure 6.1 – Ductile Failure

Another type of failure is known as *brittle fracture.* A brittle material has virtually no plastic region to its stress-strain graph; consequently, when a brittle material fractures there is virtually no plastic deformation. Figure 6.2 shows possible forms of fracture in such a situation with the surfaces of the fractured material appearing bright and granular due to the reflection of light from individual crystals.

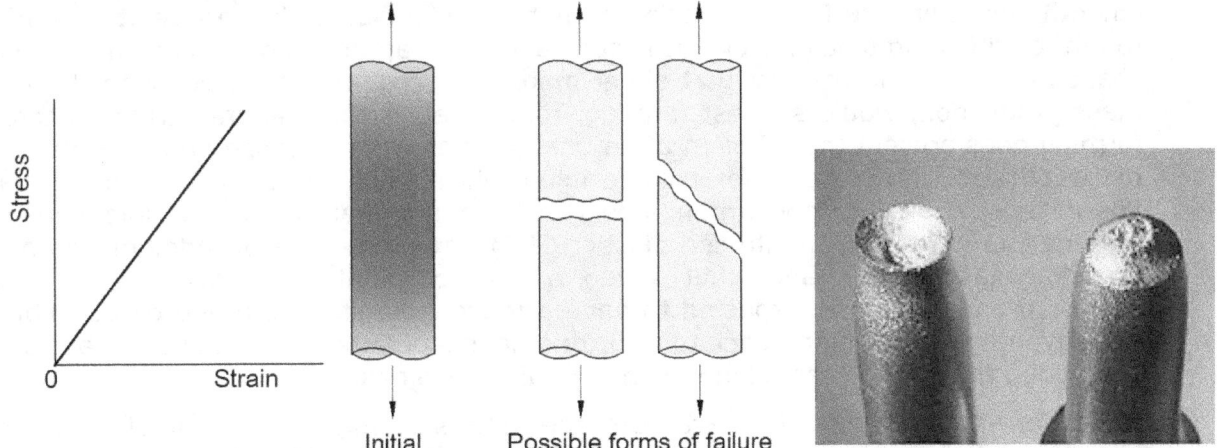

Figure 6.2 – Brittle Failure   *Figure 6.3*

Fatigue failure often starts at some point of stress concentration; the point of origin of the failure can be seen on the failed material as a smooth, flat, semicircular or elliptical region, often referred to as the nucleus.

The sample test pieces in Figure 6.3 show the cone and cup shapes resulting from the failure.

Surrounding the nucleus is a burnished zone with ribbed markings; the smooth zone is produced by the crack propagating relatively slowly through the material and the resulting fractured surfaces rubbing together during the alternating stressing of the component. When the component has become so weakened by the steadily spreading crack that it can no longer carry the load, the final abrupt fracture occurs with the region of abrupt failure having a crystalline appearance. Figure 6.4 shows the various stages in the growth of a fatigue failure.

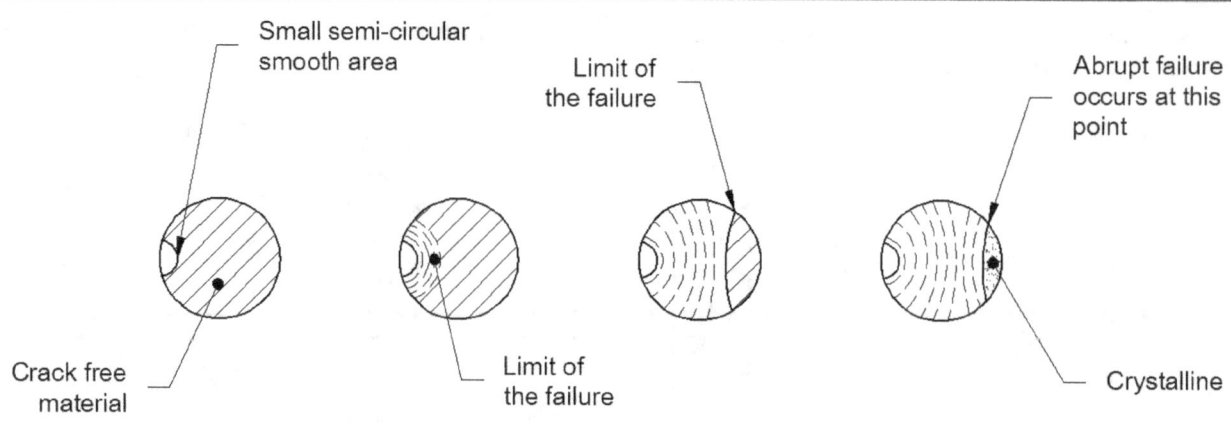

Figure 6.4 Fatigue Failure

Selection of Materials:

The following are some case studies, additional to that of the kitchen pan discussed earlier in Topic 6; they are designed to illustrate the processes involved in the selection, while there are no doubt a variety of alternative solutions to those suggested. In addition, there are likely to be many more factors involved before a selection is made.

Car Bodywork:

The functions required of car bodywork are that it protects the engine and car occupants from the weather while providing a safe environment and a pleasing appearance. The requirements for the material are that it can be formed to the shapes required, it has a smooth and shiny surface, corrosion is not too significant, in service it is sufficiently tough to withstand small knocks, it is stiff, and is cheap and can be mass produced. The shapes required and the fact that sheet material is required, together with the need for mass production, would suggest forming from sheet as the manufacturing process. Hot forming does present the problem of an unacceptable surface finish and so a material has to be chosen which allows for cold forming which indicates a highly ductile material. Possibilities would be low-carbon steels or aluminium alloys; a possible non-metallic material could be glass reinforced plastic (GRP) or carbon-fibre. In addition, both metals are reasonably tough and when given a coat of paint are reasonably resistant to corrosion and so can be expected to have a reasonable life. GRP and carbon-fibre have the advantage of springing back to their original shape after a small impact, are available in a wide range of gelcoat colours with high gloss finishes.

The stress-strain graph of low-carbon steel differs in form from that of an aluminium alloy in that the aluminium alloy shows a smooth transition from elastic to plastic deformation while the carbon steel shows some irregularities (Figure 6.5). The effect of this on the cold forming of carbon steel is to give some surface markings which do not occur with the aluminium alloy. The aluminium alloy thus has an advantage over the steel in giving a smoother surface when formed. Aluminium alloys also have the advantage that they have lower densities and so could lead to lower weight cars. The carbon steel does, however, have some advantages; it work hardens more than the aluminium alloy and so gives a harder material. In addition, the steel has a higher tensile modulus than the aluminium and resulting in a stiffer material. The great advantage, outweighing all other considerations, is that carbon steel is much cheaper than aluminium alloy; typically it is about half the price therefore the choice is a low-carbon steel over the aluminium. In practice, the steel used has less than 1% carbon.

The trend in the motor industry is to use metal for the main parts of the vehicles, and plastic or GRP for fairings and minor areas.

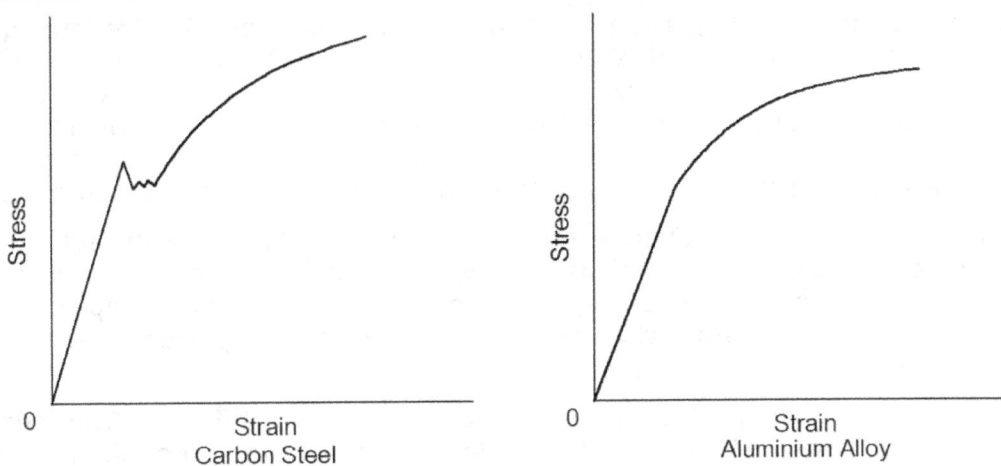

Figure 6.5 – Stress-Strain Graph for Carbon Steel and Aluminium Alloy

Tennis Racket:
The function of a tennis racket is to transmit power from the arm of the player to a tennis ball. The requirements for the frame and handle of a racket are high strength, high stiffness, low weight, toughness and ability to withstand impact loading, durability and not creeping or warping as a result of exposure to temperature or humidity changes, and ability to be processed into the required shape; another requirement which requires a little explanation is the ability to damp out vibrations. When the ball hits the strings, vibrations occur; these are then transmitted through the frame of the racket to the arm of the player. If these vibrations are not reduced in amplitude in this transmission then the elbow of the player can suffer some damage, known as tennis elbow through the elbow joint is subjected to vibration. Cost will be a factor when considering tennis rackets for the general population but is less a requirement for rackets for professional tennis players.

The requirement for high strength and low weight can be translated into a need for a high value of strength/density, i.e. specific strength. Similarly the requirement for high stiffness and low weight results in a need for a high value of modulus/density, i.e. specific modulus. Possibilities would seem to be timber, metals and composites. The following table shows values for some possible materials.

Material	Specific Strength MPa/MG^{-3}	Specific Stiffness GPa/Mgm^{-3}	Relative Toughness	Relative Vibration Dampening	Relative Coat
Timber: Ash	107	20	Good	Good	Low
Timber: Hickory	105	21	Good	Good	Low
Al-Cu Alloy 2014 Prec. Hard	15	25	Good	Poor	Medium
Al-Cu Alloy 5050 Annealed	54	27	Good	Poor	Medium
Mn Steel 120M36 Q&T	90	27	Good	Poor	Medium
Ni-Cu-Mo Steel 817M40 G&T	115	27	Good	Poor	Medium
Composite: Epoxy + 60% Carbon	890	90	Medium	Medium	High
Composite: Epoxy + 70% Glass	750	25	Medium	Medium	High

Timber has the advantages that it is tough, has good specific strength, good damping properties for vibrations and is cheap. The specific stiffness could be better. Warping could be a problem if subjected to moisture (rain); however, this can be overcome by using laminated timber, i.e. several pieces of timber with their fibres in different directions bonded together to give a laminate. The combining together of pieces of timber also gives a method by which the shape of the racket can be obtained.

Aluminium alloy has the advantages of toughness and good specific stiffness. It is more expensive than timber. A problem, however, is that it has very poor vibration damping. Aluminium can be protected against corrosion attack by damp environments by anodizing. An aluminium racket can be made by bending extruded hollow sections into the required shape.

Steels can give high specific strengths and high specific stiffness. The steels with these high strengths are likely to be comparable in price with the aluminium alloys. Problems are, however, the very poor vibration damping and the poor corrosion resistance in a damp environment unless protected by a powder coating. A steel racket can be made by bending extruded hollow sections into the required shape.

Composite materials can be made which have the advantages of very high specific strengths, very high specific stiffness, reasonable vibration damping and tolerable toughness. The major problem, however, is the high cost of such materials. A composite racket can be made by injection moulding a melt of a polymer containing carbon fibres into a racket shaped mould. The process would give a racket with a solid composite for the frame and handle. The procedure that can then be adopted to improve the properties is, while the racket is still in the mould and only the outer skin of the composite has solidified, to pour out the liquid core so that when the racket solidifies there is a hollow tube. The tube can then be filled with polyurethane foam to improve the vibration damping of the racket.

In comparing the above materials, the composite material racket gives the best properties but is considerably more expensive than the others. It therefore is more likely to be used by the professional tennis player. For cheapness and properties, timber is probably the best material based on economics, followed by aluminium alloys with steel being the worst. The modern society would most like select aluminium due to the perception of a modern material and brighter range of colours.

Small Components for Toys:
Consider small components such as the wheels for, say, a small model toy car for use by a child; the functions required of the wheels are that they are safe and rotate on their axles. The materials thus need to be non-toxic, reasonably tough, not easily deformed by knocks, not brittle and cheap. Before considering possible materials, AS/NZ 8124-2008 should be consulted and covers the safety aspects related to mechanical and physical properties; particularly the use of arsenic, barium, cadmium, chromium, lead, mercury and selenium from toy materials.

The products are required to be cheap when produced in relatively large quantities and the products themselves are rather small. In the case of metals the obvious process is die casting. Though the initial cost of the die is high, a large number of components can be produced from one die and so the cost per component becomes relatively low. In the case of polymers the obvious choice is injection moulding resulting in a high die cost but large numbers of components can be produced from one die and hence the cost per component can be low. Both processes give good surface finish and dimensional accuracy.

In the case of metals, die casting limits the choice to those with relatively low melting points, i.e. aluminium, magnesium, zinc, lead and tin alloys. Table 6.3 shows relevant properties of these materials. Safety considerations ban the use of lead. Aluminium, magnesium and zinc are comparable in cost, with zinc tending to have the lower cost per unit weight. Tin is more expensive than the alloys. Zinc has a lower melting point than aluminium or magnesium and in the as-cast condition has the highest tensile strength. Zinc would seem to be the best metal choice for the product.

Alloy	Density Mg m^{-3}	Melting Point °C	Strength MPa
Aluminium	2.7	600	150
Lead	11.3	320	20
Magnesium	1.8	520	150
Tin	7.3	230	12
Zinc	6.7	380	280

Zinc is very widely used for die casting. Its low melting point and excellent fluidity make it one of the easiest metals to cast. Small parts of complex shape and thin wall sections can be produced. Zinc alloys have relatively good mechanical properties and can be electroplated.

Polymers are a possible alternative to metals. Since the forming method is to be injection moulding, the materials are restricted to thermoplastics.

The choice is then of polymers which are relatively stiff. The following table compare the properties of possible polymers to those of zinc. The mechanical properties of the zinc alloy are superior to those of thermoplastics, it having higher strength, higher tensile modulus, and being tougher and more resistant to fatigue and creep. Where light weight is required then polymers have the advantage, having densities of the order of one-sixth that of zinc. Where coloured surfaces are required then polymers have the advantage since pigments can be incorporated into the polymer mix; however, if electroplating is required then zinc has the advantage. On cost per unit weight then zinc is cheaper, but the interest is likely to be cost per unit volume and on this basis polymers are likely to be cheaper.

Alloy	Strength MPa	Modulus GPa	Density Mg m^{-3}	Relative Cost m^3
ABS	50	2.3	1.02 – 1.07	1
Nylon 6	60	3.2	1.13 – 1.14	2
Polycarbonate	65	2.3	1.2	2
Zinc Alloy	280	103	6.7	3

On the basis of the above considerations, it is likely that zinc would be used where fine detail and colour of surface is not a requirement but a metallic surface as a result of electroplating is required. Otherwise polymers, possibly ABS, would be used.

MEM30007A - Select common engineering materials
Topic 6 - Selection of Materials

Review Problems:

MEM30007-RQ-06

1. Make reasoned proposals for materials and processes for the following products:
   a) domestic window catches,
   b) structural I-beams for use in building construction,
   c) rainwater gutters and drainpipes,
   d) a domestic washing-up bowl,
   e) pipe through which sea water can be pumped,
   f) small fan in a vacuum cleaner,
   g) the lenses for the rear lights of cars,
   h) a camshaft for a car,
   i) casing for a hand-held power tool,
   j) blades for a lawn mower.

2. Investigate the materials used with the following products and give reasons why they might have been chosen in preference to others and the processes that might have been used:
   a) the casing for mains electric plugs,
   b) spades,
   c) domestic cold and hot water pipes
   d) the casing for the body of a vacuum cleaner,
   e) joists to support floors in a small house.

Topic 7 – Safety Parameters:

Required Skills:
- Recognize the responsibility of the individual to others and the organization in the need to work within required safety parameters, i.e. Health and Safety at Work.
- Assess the risks in the workplace associated with the use, handling, processing, storage and disposal of materials, i.e. with the Control of Substances Hazardous to Health.

Required Knowledge:
- Health and safety regulations and procedures in the workplace.
- Types of protective clothing.
- Reading graphs, tables and charts.
- Testing procedures.
- Application of types of materials.

Health and safety at work
Safety in a company is everyone's responsibility. While there may be some employees, such as safety officers or safety representatives, who have special responsibilities, all the employees and the employers have responsibilities. The standard governing with Health and Safety at Work is AS 1470-198. The code lays out the legal requirements and provides a comprehensive and integrated system of law in relation to the health, safety and welfare of people at work and for the protection of the general public against risks to health and safety arising out of, or in connection with, the activities of persons at work. AS 1470 consists of the following sections.
- Section 1 – Scope and General
- Section 2 – Responsibilities
- Section 3 – Principles and Techniques
- Section 4 – Strategies for Control
- Section 5 – Organisational Arrangements
- Section 6 – Personnel
- Section 7 – Occupational Health Services
- Section 8 – Workplaces and the Working Environment
- Section 9 – Machinery, Plant and Equipment
- Section 10 – Storage and Handling of Material
- Section 11 – Control of Harmful Chemicals
- Section 12 – Fire and Explosion
- Section 13 – Vehicle Operation
- Section 14 – Personal Protection Equipment

Employer's Responsibilities
Under the code, the general duties are laid down for employers. These are to ensure, as far as is reasonably practicable, the health, safety and welfare at work of all employees extending to:

a) Setting out in writing their policy and arrangements in the field of occupational health and safety and bringing this information to the notice of employees in language they readily understand.

b) Controlling or eliminating hazards at work and, where appropriate, the provision of protective clothing.

c) Providing adequate supervision of work, of work practices and of the application and use of occupational health and safety measures.

d) Giving necessary instructions and training, taking account of the functions and capacities of employees.

e) Providing, where necessary, for measures to deal with emergencies and accidents, including adequate first-aid arrangements, and for rehabilitation.

f) Verifying the implementation of applicable standards on occupational health and safety regularly, for instance by environmental monitoring, and undertaking systematic audits from time to time.

g) Keeping such records relevant to occupational health and safety and the working environment as are considered necessary by the competent authority or authorities. These might include records of all notifiable accidents and personal damage which arise in the course of or in connection with work, records of authorization and exemptions under laws or regulations in the field and any conditions to which they may be subject, certificates relating to supervision of the health of employees in the enterprise, and data concerning exposure to specified substances and agents.

Whenever two or more employers or self-employed persons engage in activities simultaneously at one workplace, they each have a responsibility to coordinate and conduct those activities in such a way as will ensure, so far as is reasonably practicable, that nothing about the manner in which each business is conducted makes it unsafe or a risk to the health or safety of not only their own employees but also persons not in their employment.

Employee's Responsibilities:
The code lays down the responsibility for the employees in the course of performing their work including:
a) Taking a reasonable care for their own safety and that of other persons who may be affected by their acts or omissions;
b) cooperating in the fulfilment of the obligations placed upon their employer;
c) complying with instructions given for their own health and safety and those of others and with health and safety procedures;
d) use safety devices and protective equipment correctly and not render them inoperative;
e) immediately reporting to their immediate supervisor any situation which they have reason to believe could present a hazard;
f) reporting any accident or injury to health which arises in the course of or in connection with their work.

Government Responsibilities:
Governments have a responsibility to:
a) promote improvements in the quality of the working environment, including safe and healthy work practices and workplaces;
b) consult with employer and employee organizations prior to the introduction of legislation;
c) enact legislation concerning occupational health and safety and the working environment which is consistently and effectively enforced by an adequate and appropriate system of inspection and which includes penalties for breaches of the laws and regulations;

d) cooperate in the adoption, so far as is practicable, of uniform standards for safe and healthy working environments and work practices;
e) provide guidance to employers and employees to help them comply with legal obligations;
f) co-ordinate occupational health and safety activity;
g) undertake and stimulate education, research, and training; monitor the health of people in the workplace, including the incidence and prevalence of work-related injuries and disease; and
h) collect, compile, evaluate, and disseminate information on:
   i. working environment;
   ii. occupational health and safety statistics, principles, practices and standards; and
   iii. occupational hazards and the means to eliminate or control them.

Trade Union Responsibilities:
Trade unions and employee associations have a responsibility to:
a) identify the health and safety problems for occupations of their members;
b) communicate these problems to their members, individual enterprises and industry associations.
c) cooperate with and assist government, employer, and industry associations in developing preventive strategies in occupational health and safety.

Designers, Manufacturers, Importers and Suppliers Responsibilities:
Designers, manufacturers, importers, and suppliers have a responsibility to ensure, so far as is reasonably practicable, that:
a) plant, machinery, or equipment is designed, tested, and installed or constructed so as to be free from avoidable risks to health or safety when not misused;
b) any substance for use at work is free from avoidable risks to health or safety when properly used; and
c) adequate information is made available about the correct installation and use of machinery and equipment or use of substances and about any condition necessary to ensure that it will be free from avoidable risk to health or safety when properly used.

Industry Association Responsibilities:
Industry associations have a responsibility to:
a) identify health and safety problems within the enterprises of their members, or in enterprises within the same industry or utilizing similar substances or processes;
b) communicate these problems to their members, and other enterprises within the same industry; and
c) assist in the identification of countermeasures and to communicate such information to members and other enterprises within the same industry.

Safe Work Systems:
Employers have the duty that they should provide systems of work that are, as far as is reasonably practicable, safe and without risks to health; this requires the establishment of safe work systems and a consideration of such factors as:

Health:
Meaning a consideration of the hazards to health of exposure to certain chemicals, of what happens when chemicals are spilt, etc.

Safety:
Meaning a consideration of the hazards that can occur when a machine or its guard fails, or an operator chooses to do a job in a different way, etc.

MEM30007A - Select common engineering materials
Topic 7 - Safety Parameters

Safe Operating Procedures:
A written Safe Operating Procedure (SOP) should be provided near all equipment to assist in reducing workplace hazards. The SOP states exactly the procedures to be followed when using the equipment by a responsible person.

The Needs of Individuals:
The may include protective clothing and equipment, seating and working space being considered in relation to the needs of each individual; there are also such considerations as whether they are able to understand safety instructions and can work safely if under medication, have some handicap, etc.

Maintenance:
In order to ensure continuing safety, equipment, buildings and plant need a planned maintenance schedule prepared and adhered to.

Monitoring:
The system needs checking to ensure that rules and precautions are being followed and continues to deal with the risks. New hazards may be found to be introduced by changes in staff, materials, equipment, etc.

Protective Clothing and Equipment:
Wherever possible, employers should eliminate or control risks so that protective clothing and equipment may not be required. Some jobs, however, require protective clothing or equipment specified by law. The following is an indication of some of the hazards and protection that might be used:

	Hazards	Protection
Eyes:	Chemical or metal splashes, dust, projectiles, gas, radiation.	Spectacles, goggles, face screens, helmets.
Head and Neck:	Falling objects, bumping head, hair entanglement, chemicals	Helmets, bump caps, helmets, caps, skull-caps.
Hearing:	Impact noise, high levels of sound, high- and low-frequency sound.	Earplugs, muffs.
Hands and Arms:	Abrasion, cuts, temperature extremes, chemicals, skin infections, vibration, electric shock.	Gloves, gauntlets, armlets, wrist cuffs.
Feet and Legs:	Wet, slipping, cuts, abrasion falling objects, heavy pressures, metal and chemical splashes.	Safety boots, gaiters, leggings.
Respiratory:	Toxic and harmful dusts, gases and vapours, microorganisms.	Disposable respirators, mask respirators, fresh air hose equipment.
The Body:	Heat, cold, weather, chemical or metal splashes, contaminated, , impact, dust, entanglement of clothing.	Overalls, boiler suits, warehouse coats, donkey jackets, aprons, specialist protective clothing.

Accidents and Emergencies:
A plan for dealing with emergencies should be formulated, whether they are simple or major incidents. People need to be told what might happen, how the alarm will be raised, what to do, where to go, who will be in control, and what essential actions need

to be taken, e.g. closing down plant. The employees need training in emergency procedures. Access-ways need to be kept clear for emergency services and escape routes. Fire-fighting equipment, electrical isolators and shut-off valves need to be clearly labelled. Emergency equipment needs testing regularly.

After an accident or a serious incident, the immediate emergency should be dealt with, any injuries treated and the premises or plant made safe. Any injuries should be recorded in an accident book and if applicable, the incident should be reported to authority responsible for workplace safety.

A Safe and Healthy Environment:

The employer is charged with the provision and maintenance of a working environment for the employees that is, as far as is practicable, safe, without risks to health and adequate as regards facilities and arrangement for their welfare at work. The employer has thus to be concerned with such facilities as hygiene and welfare; cleanliness; floors and gangways being kept clean, dry and not slippery; seats, machine controls, instruments and tools being designed for the best control, use and posture; the place of work being safe with adequate space for easy movement and safe machine adjustment, no tripping hazards, emergency provisions, etc.; lighting that gives good general illumination with no glare, adequate emergency lighting, etc.; and a comfortable environment with a suitable working temperature, good ventilation, acceptable noise levels, etc.

Health hazards can arise in a number of ways since people at work are exposed to a wide range of substances. Australian Standard AS 2714 deals with the storage and handling of hazardous chemical substances while AS 1470 outlines the reduction of risks to health and safety from hazardous substances emitted by machinery.

Hazardous substances include:
- those listed in an approved list as dangerous and for which the general indication of the nature of risk is specified as very toxic, toxic, harmful, corrosive or irritant;
- for which a maximum exposure limit has been specified or for which there is an occupational exposure standard;
- any microorganism hazardous to health;
- dust of any kind when present at a substantial concentration in air;
- any other substance which could create a hazard to health comparable to those listed above.

Employers are required to:
- Assess the risk of exposure to hazardous substances and the precautions which should be taken to prevent exposure, or if this is not reasonably practicable, to achieve adequate control.
- Introduce appropriate control measures to prevent or control the risk.
- Ensure that the control measures are used.
- Ensure that the control measures are properly maintained and periodically examined.
- Where necessary, the exposure of workers should be monitored. '
- Where necessary, carry out health surveillance of workers.
- Inform, instruct and train employees and their representatives of the risks involved, the precautions to be taken and the results of monitoring.

As an indication of the types of hazards needing to be controlled, consider the processing of plastics. Moulding and extrusion machines have automatic alarm systems to limit overheating of polymers; these may not always be correctly set. Heating thermoplastic materials to beyond their processing range causes them to decompose and produce flames. The normal processing of polymers may also give rise to fumes and in some cases these fumes can be toxic; in addition, there may be dust hazards during the supply of dry materials to the feed units of the machines. The hazards posed by such fumes and dust needs to be assessed and guarded against.

MEM30007A - Select common engineering materials
Topic 7 - Safety Parameters

Review Problems:

MEM30007-RQ-07

1. What are the general duties imposed by the Health and Safety at Work code on (a) employers and (b) employees?

2. Explain the functions of (a) improvement notices and (b) prohibition Notices

3. Obtain one of the following Health and Safety Executive booklets and produce a paper outlining the consequences of such guidance on operations.
   a) Safety devices for hand- and foot-operated presses,
   b) Safety in drop forging hammers,
   c) Safety in the stacking of materials.

4. Discuss the following situations and consider what actions should be taken or what consequences could occur:
   a) A worker removes the safety guards from a machine because they reduce the number of items he can produce and hence his wages, which are based on piece-rate.
   b) The company is short of storage space for a large consignment of goods that has just arrived and so they temporarily stack them in the passage ways leading to the main exit from the factory floor.
   c) A worker is just about to become married and in celebration fellow workers set off the fire extinguishers.
   d) An inspector wants to go into the tool room but the management tell him that they will not allow it because the work there is commercially highly secret and he might tell their competitors.
   e) An inspector wants to talk to a worker about his machine but the worker refuses to talk to him.
   f) A worker is injured and as a result is off work for a month. The employers pay him or her during that time but take no other action.

5. What protective clothing and equipment might be suitable in the following situations?
   a) Chemicals may splash into the eyes.
   b) A press is very noisy during operation.
   c) A storekeeper has to move sheets of metal.

MEM30007A - Select common engineering materials
Review Question Answers

Answers:

Topic 1:
1. These might include (a) ease of forming in one piece, easily cleaned, strain resistant, waterproof; (b) stiff, strong, cheap; (c) leakproof, suitable for hot liquids, cheap, not easily broken; (d) good conductor, flexible; (e) cheap to manufacture, wear resistant during handling, stiff; (f) withstands changing forces, stiff strong, withstands impact forces; (g) attracting appearance, cheap to form.
2. (a) Stainless steel; (b) Timber; (c) China - a ceramic; (d) Copper; (e) alloys of copper (cupronickel or bronze depending on the colour of the coins; (f) steel; (g) Plastic, e.g. ABS.
3. (a) Modulus; (b) Ductility, percentage elongation; (c) Fracture toughness; (d) Strength; (e) Electrical resistivity/conductivity; (f) thermal conductivity; (g) Corrosive properties.
4. Strong and brittle.
5. Strong and tough.
6. 20 MPa.
7. 0.67%
8. 50 Kn
9. 12%
10. 50 Kn
11. The bronze is stronger and more ductile.
12. Stronger in compression, brittle.
13. Ductile above 0°C, brittle below.
14. Reasonably good.
15. Very low resistivity, of the order of 10^{-8} W m.
16. 0.0125 W m.
17. Thermoplastics: flexible, soft, can be formed by heating; thermosets; rigid, hard, cannot be formed by heating.
18. Brittle, must not be subject to sudden forces or sudden changes in temperature.
19. 128 MPa/Mg m^{-3}, 33 MPs/Mg m^{-3}, 0.78 $/MPa.

Topic 2:
1. Cast iron 0.014 GPa/kg m-\l alloy 0.027 GPa/kg m-^, PVC 0.002 GPa/kg m-3
2. Steel 220 GPa, Al alloy 71 GPa, Polypropylene 1-2 GPa, Composite 20 GPa
3. 120 MPa/Mg m-^
4. 1.4-3.1 GPa, in the high range of modulus values for plastics
5. 470-570 MPa, 170-280 MPa, 18-35%
6. Silicon, 2%
7. 5052
8. Polyacetal
9. 226M44

Topic 3:
1. (a) 61 GPa, (b) 380 MPa
2. (a) 10.8 MPa, (b) 1.1 GPa
3. (a) 660 MPa, (b) 425 MPa, (c) 200 GPa
4. 31.1%
5. (a) 480 MPa, (b) 167 GPa
6. 300 MPa, 280 MPa
7. 2.5 GPa, 80 MPa
8. See Figure A. 1
9. Stronger and less ductile
10. (a) Titanium alloy, (b) nickel alloy
11. Cellulose acetate
12. Becoming more ductile
13. Becoming more brittle
14. As the temperature drops becoming more brittle
15. Becoming more ductile

MEM30007A - Select common engineering materials
Review Question Answers

16. HV 198
17. HV 275
18. HV 71
19. HB 217
20. HB 57
21. After exposure breakdown voltage decreases
22. Ni-Cr alloy most corrosion resistant
23. Increasing carbon reduces oxidation, increasing chromium reduces oxidation.
24. Industrial pollutants more damaging than marine conditions, with rural surroundings being least corrosive.
25. (a) Hardness test, (b) impact test, (c) tensile test for the modulus of elasticity, (d) impact, or tensile or hardness, test, (e) bend test.

Topic 4:

1. A crystal within a metal, i.e. a region of orderly packing of atoms.
2. A mixture of two or more elements, e.g. iron and carbon in steel.
3. Ferrous alloys have iron as the main constituent, a non-ferrous alloy a metal other than iron.
4. An array of grains, i.e. crystals, within which there are orderly arrays of atoms J
5. More ductile, the bigger the grains
6. Elongated grains give different properties in the directions of the grains compared with at right angles
7. Grains become elongated and distorted with an increasing number of dislocations. The tensile strength and hardness increases, the ductility decreases.
8. 8 See Figure 4.7
9. (a) Large grain, few dislocations, (b) small grain, many dislocations.
10. Increase in dislocations and hence an increase in yield strength, tensile strength and hardness but a decrease in ductility.
11. See Figure 4.19
12. See Figure 4.19. Above the recrystallization temperature no work hardening occurs
13. Grains elongated in direction of rolling.
14. Cold rolled has distorted grains, is work hardened and directionality of properties. Hot rolled has large grains, is ductile and has surface oxide layers.
15. Greater than 400°C, probably about 500°C.
16. (a) More distortion and dislocations, increased hardness and brittleness, (b) up to 300°C, (c) above 300°C.
17. (a) About 110 HV, (b) about 30 HV, (c) roll, anneal, roll, anneal, roll so that the final rolling gives less than 10% reduction.
18. See Figures 4.22 and 4.23. The greater the crystallinity, the greater the density, melting point and strength.
19. LDPE is a branched polymer with less crystallization than HDPE which is a linear polymer. See Table 4.1
20. (a) To protect against UV and resist deterioration, (b) to make more flexible, (c) to reduce cost, increase perhaps strength, impact strength, resistivity, or reduce friction.
21. See Table 4.3
22. Makes it more rigid.
23. See Figure 4.31 and associated text.
24. 36.4 GPa, in direction of fibres.
25. 182.2 GPa
26. 205 GPa
27. The long fibres give directionality of properties and a greater improvement in strength and modulus than random fibres, which give no directionality.

Topic 5:

1. Better surface finish with cold drawing. The heating anneals the material to make it soft and ductile.
2. Oriented distorted grains and hence work hardened with directionality of properties.
3. Die casting.
4. Slow cooling gives high degree of crystallinity.

5. Molecules aligned along the direction of the extrusion.
6. Molecules aligned with the direction in which the material was stretched.
7. Molecules lined up along the length of the bag.
8. Gives molecular alignment and improves the strength.
9. (a) Recrystallization and grain growth, ductility improves; (b) martensite forms as carbon atoms become trapped, increase in hardness; (c) some carbon atoms diffuse out of martensite, increase in ductility; (d) fine particles slowly move out of quenched material into dislocations and grain boundaries, increase in hardness; (e) surface layers become martensitic, increase in hardness, (f) carbon diffuses into outer layers, increase in surface hardness.
10. Annealing gives grain growth and a soft structure; quenching gives martensitic structure and increase in hardness, strength and brittleness; tempering allowing carbon atoms to diffuse out of the martensite and so reduce the structural distortion and hence brittleness.
11. Annealing gives grain growth and a soft, weak structure; work hardening distorts grains and introduces dislocations with the result that the material is stronger, harder and more brittle.
12. Annealing gives grain growth and a soft, weak structure, precipitation hardening causes fine particles to become lodged at grain boundaries and dislocations, hence increasing the strength, hardness and brittleness.
13. The hammer head is forged and then the striking surface is surface hardened, possibly by flame hardening followed by tempering.
14. The blade has to be tough with the teeth hard. Cast ingots are hot rolled, then blade-size strips cut out and teeth machined. Surface hardening, flame hardening, is then used for the teeth followed by tempering in order to achieve the required hardness and not too brittle a state for the teeth.

Topic 6:
1. 1 (a) Aluminium alloy, e.g. LM9, die cast; (b) carbon steel, about 0.3% carbon, or low-alloy steel such as a chromium-molybdenum steel, hot rolling; (c) unplasticized PVC, extrusion; (d) high-density polyethylene, injection moulding; (e) high-density polyethylene, extrusion; (f) nylon, injection moulding or die cast aluminium alloy, e.g. LM9; (g) acrylic (polymethyl methacrylate), injection moulding; (h) medium-carbon steel, about 0.4% carbon, or low-alloy steel such as 530M40, forged, quenched and tempered; (i) polypropylene, injection moulded or aluminium alloy, e.g. LM6, die cast; (j) ABS, injection moulding
2. 2 (a) Thermoset, e.g. urea formaldehyde, moulding; (b) medium-carbon steel or stainless steel, forging; (c) copper, e.g. CI06, extrusion; (d) ABS, injection moulding
3. 3 See the British Standards

Topic 7:
1. See Sections 7.1.1 and 7.1.2
2. (a) Improvements are required, (b) the process or equipment is prohibited from being used
3. See the HSE booklets.
4. (a) Contravenes HSW Act duties imposed on employees, (b) employers are failing to ensure the welfare of employees in an accident or emergency, (c) contravenes HSW Act duties imposed on employees, (d) the inspector has the right, (e) the worker cannot refuse, (f) should report it.
5. Could be: (a) spectacles, goggles, face screens; (b) earplugs or muffs; (c) safety boots, gloves.

www.ingramcontent.com/pod-product-compliance
Lightning Source LLC
Chambersburg PA
CBHW081047170526
45158CB00006B/1885